ONE HUNDRED PHYSICS VISUALIZATIONS USING **MATLAB**

Second Edition

Other World Scientific Titles by the Author

Physics at Fermilab in the 1990's: Proceedings of the Workshop
ISBN: 978-981-02-0103-6

Lectures in Particle Physics
ISBN: 978-981-02-1682-5
ISBN: 978-981-02-1683-2 (pbk)

At the Leading Edge: The ATLAS and CMS LHC Experiments
ISBN: 978-981-4277-61-7
ISBN: 978-981-4304-67-2 (pbk)

One Hundred Physics Visualizations Using MATLAB
(With DVD-ROM)
ISBN: 978-981-4518-43-7
ISBN: 978-981-4518-44-4 (pbk)

More Physics with MATLAB
With Companion Media Pack
ISBN: 978-981-4623-93-3
ISBN: 978-981-4623-94-0 (pbk)

Cosmology with MATLAB
With Companion Media Pack
ISBN: 978-981-310-839-4
ISBN: 978-981-310-840-0 (pbk)

Beams and Accelerators with MATLAB
With Companion Media Pack
ISBN: 978-981-323-746-9

Stars and Space with MATLAB Apps
With Companion Media Pack
ISBN: 978-981-12-1602-2
ISBN: 978-981-12-1635-0 (pbk)

The Physics of Experiment Instrumentation Using MATLAB Apps
With Companion Media Pack
ISBN: 978-981-12-3243-5
ISBN: 978-981-12-3383-8 (pbk)

One Hundred Physics Visualizations Using MATLAB
Second Edition
ISBN: 978-981-12-9561-4

Cosmology with MATLAB
Revised with MATLAB Live Scripts
Revised Edition
ISBN: 978-981-98-0103-9

ONE HUNDRED PHYSICS VISUALIZATIONS USING **MATLAB**

Second Edition

Dan Green
Fermi National Accelerator Laboratory, USA

World Scientific

NEW JERSEY • LONDON • SINGAPORE • BEIJING • SHANGHAI • HONG KONG • TAIPEI • CHENNAI • TOKYO

Published by

World Scientific Publishing Co. Pte. Ltd.
5 Toh Tuck Link, Singapore 596224
USA office: 27 Warren Street, Suite 401-402, Hackensack, NJ 07601
UK office: 57 Shelton Street, Covent Garden, London WC2H 9HE

Library of Congress Control Number: 2024943402

British Library Cataloguing-in-Publication Data
A catalogue record for this book is available from the British Library.

ONE HUNDRED PHYSICS VISUALIZATIONS USING MATLAB
Second Edition

ISBN 978-981-12-9561-4 (hardcover)
ISBN 978-981-12-9562-1 (ebook for institutions)
ISBN 978-981-12-9563-8 (ebook for individuals)

For any available supplementary material, please visit
https://www.worldscientific.com/worldscibooks/10.1142/13913#t=suppl

Desk Editor: Carmen Teo Bin Jie

Typeset by Stallion Press
Email: enquiries@stallionpress.com

Preface

"The soul should always stand ajar, ready to welcome the ecstatic experience" — Emily Dickinson

"The effort to understand the Universe is one of the very few things that lifts human life above the level of farce, and gives it some of the grace of tragedy." — Steven Weinberg

This volume is now more than a decade old, and much has transpired during that time. The MATLAB tools have evolved from scripts, to Apps and at present to Live code. The Live package is preferred because it combines text and equations with MATLAB code all in a single site. The results of that code, formerly shown separately, also appear in line and in this way the user can vary the parameters of the specific problem and explore immediately how the solutions vary in response. For this reason the Live scheme is used exclusively in this edition.

The physics landscape has also evolved in the last decade. These changes have been reflected in the problems which are explored in this volume. More emphasis on astrophysics and cosmology reflects the advances in physics over the last decade. The Nobel prize in 2006 rewarded the discovery of small perturbation in temperature, at the parts per million level of the extreme isotropy of the Cosmic Microwave Background (CMB). The basic isotropy is now thought to indicate a period of rapid expansion of the Universe, called "inflation".

The 2011 Nobel prize was for the observation, using supernovae as "standard candles", of the accelerating expansion of the universe. This expansion is now ascribed to the existence of "dark energy". In 2013 the prize was awarded for the discovery of the Higgs boson, a fundamental scaler, and the first and only such fundamental particle. Indeed, the Higgs boson has the quantum numbers of the vacuum, as would the driver for "inflation".

In 2017 the prize went for the observation of gravitational radiation using gravity wave detectors. In 2019 the prize was awarded for the explication of the structures in the CMB and the subsequent emergence of the cosmic "standard model" where the Universe is composed of matter, photons, dark matter and dark energy. In 2020, the prize was given for the exposition of the nature of the singularities of General Relativity (GR), black holes.

These awards have altered the choices made in the hundred problems that are addressed in this revised volume. The problems are a bit more complex than those of the first addition, but the content is more topical and the MATLAB tools are up to the work asked of the Live scripts.

Introduction

"Computers are useless. They can only give you answers"
Pablo Picasso

"The purpose of computing is insight, not numbers"
Richard Hamming

There are only a very few solvable problems in physics. They are extremely useful because the equations for the solutions can be plotted and the parameters defining the solutions can be varied in order to explore the dependence of the solutions on the variables of the problem. In that way the student can build up an intuition about the Kepler problem for example.

For the other problems, numerical methods are needed and the computation becomes somewhat cumbersome. As a result, it is more difficult to vary the inputs to the problem numerically rather than symbolically and develop an intuition about the dependence of the solution on those parameters. In particular time development is often obscure and the "movies" employed in this volume can be a welcome tool in improving physical intuition.

Indeed, the aim of this book is to use the ensemble of symbolic and numeric tools available in the MATLAB suite of programs to illustrate representative numerical solutions to about one hundred problems spanning several physics topics. The student typically works through the demonstration and alters the inputs using a Live script. This tool combines text, equations and solutions as figures and numerical or symbolic output all in a single Live script. The parametric variation is enabled using the Live tools of "edit fields", "dropdown menus" and numerical "sliders".

MATLAB is a good vehicle for the computational tasks. It has a compiler, editor and debugger which are very useful and user friendly. The HELP utility is very extensive as is the complete documentation set. The MATLAB language is similar to a modern C++ or Python language. Indeed code conversion between MATLAB and Python is available which expands the pool of possible users of this text. MATLAB is a vector/matrix language which makes coding simple. Data is easily imported and exported in a variety of formats.

MATLAB contains many special functions. Matrices and linear algebra are covered well. Curve fitting, polynomials and fast Fourier transforms are supplied. Numerical integration packages are available. Differential equations, symbolic, ordinary and partial, as well as numerical solutions are available for both initial value and boundary value versions.

MATLAB also has symbolic mathematics. Within that package, calculus, linear algebra, algebraic equations and differential equations are covered. It is easy to combine a symbolic treatment of a problem with a numerical display of the solution when that is desirable. In this way converting from symbols to numbers is easily achieved.

Finally, and very importantly, MATLAB has an extensive suite of display packages. One can make bar, pie, histogram and simple data plots. There are several contour and surface plots which are possible. The time evolution of solutions can be made into "movies" that illustrate the speed of a process. These extensive visualization tools are crucial in that the student can plot, vary parameters and then re-plot. There are two and three dimensional plots of all types available.

The aim of using these tools is to create intuition, not to solve a specific problem or to complete a specific number crunching exercise. The equations used will typically not be derived, since many academic texts are available that do just that. Indeed, the aim of the text is not to teach physics but to give the user a sense of how the solutions of a given physics problem depend on the parameters of that problem and to show the connections between, say, wave optics and quantum mechanics by approaching a wide set of topics in physics and noting commonalities.

MATLAB has evolved considerably since the publication of the text a decade ago. Initially MATLAB used "scripts" which functioned largely through the "Command Window" where input was made via keyboard and output was printed out or plotted in distinct figures. The evolution to Apps made tools available to vary parameters and study the results using plots. However there was a distinct App Editor view and a separate Code View;

code and output were disconnected with 2 distinct views. This text uses the "Live" coding option. The tool is all in one, unifying code, output plots, text, equations, and parameter variation tools. Using the tools provided here as examples, the reader can write scripts of their own choosing, making the scripts an open ended gateway to further exploration.

The topics for this book start with mathematics, which is basically nineteenth century. The next chapters explore sample problems in mechanics and electromagnetism. These two topics are the main ones for that time. Then waves and optics and fluids and gases are looked at, rounding out the nineteenth century. The following chapter moves into the twentieth century with examples from quantum mechanics. Continuing with the time ordering of topics, special and general relativity is covered in the next chapter. The last chapter is on astrophysics and cosmology, which tracks more recent developments in physics.

Typically the numerical exercises are performed in the standard MKS system of units, which is the international standard. However, occasionally cgs units are used, typically in area of physics where it is normally the default. That is a fact of life, and the user is expected to converted between the systems as needed. In particular, the table of most common physical constant shown in Appendix 9.1 uses MKS units.

Many very interesting topics in physics are not covered since only 100 are explored. However, the reader should be able to write scripts that would examine topics of specific interest to them. References are not spelled out since the great improvement in web browser search engines makes such a list quite redundant.

Contents

Chapter 1

Symbolic Mathematics and Math Tools

"If people do not believe that mathematics is simple, it is only because they do not realize how complicated life is." John von Neumann

"Pure mathematics is, in its way, the poetry of logical ideas." Albert Einstein

"The chief forms of beauty are order and symmetry and definiteness, which the mathematical sciences demonstrate in a special degree". Aristotle

1.1 Startup

First install and open MATLAB. Opening MATLAB brings up the Command Window, shown below. The prompt is $\gg$.

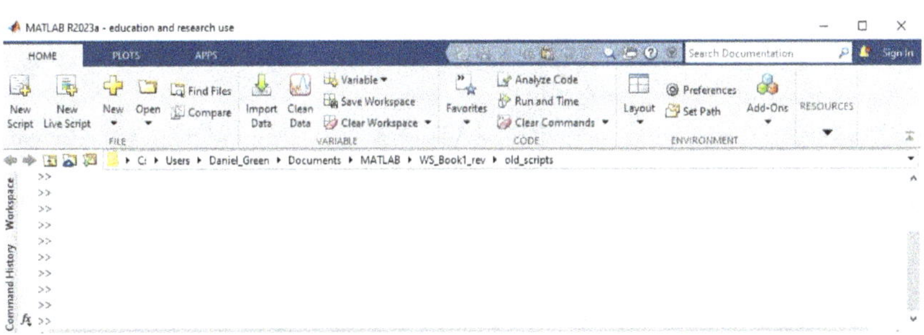

Figure 1.1: MATLAB Command Window.

Other windows, such as the Workspace window that shows all active variables and the Command History that shows all past session commands, are docked in this view with the view chosen using the Layout. The Editor opens with the "New" or "Open" tab for new or existing code. The Search Documentation provides a complete search of the extensive MATLAB documentation and often gives illustrative examples. For example, a search for documentation yields:

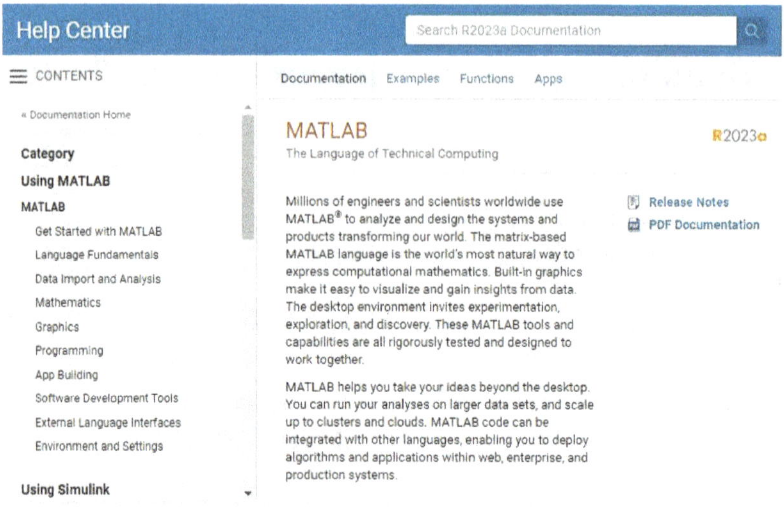

Figure 1.2: Tutorial on Using MATLAB from the Search Documentation window.

The reader is not limited to MATLAB. There are many home institutions with a site-wide MATLAB license which gives students access to the scripts. One is not limited to the language either. For example, the free software package, PYTHON, can be used since translation Apps of MATLAB scripts are also freely available.

Many tools are included with MATLAB. That fact drives the decision to use MATLAB tools to avoid unnecessary algebraic tedium in this text. Most problems are solved explicitly using the symbolic math tools. In general numerical results use the extensive MATLAB library of special functions and of numeric solvers of ordinary and differential equations.

1.2 Onramp Course

To refresh/get started the MATLAB "Onramp" is also very useful. Login to Mathworks and then search for "Onramp" for a good MATLAB introduction. The "resources" tab also has useful tutorials.

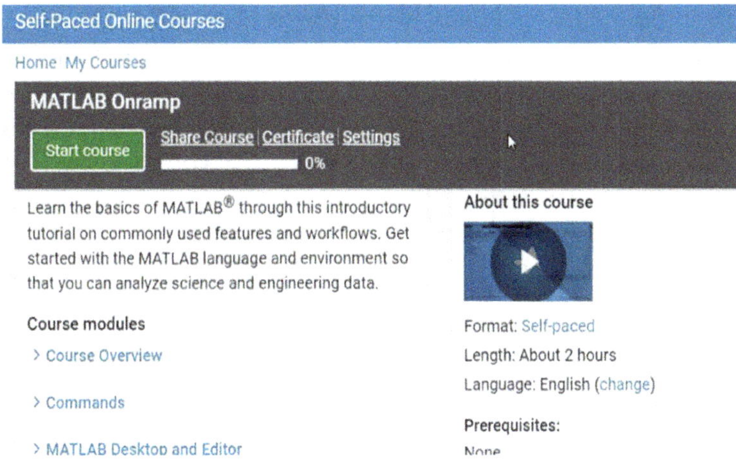

Full list of Onramp topics:

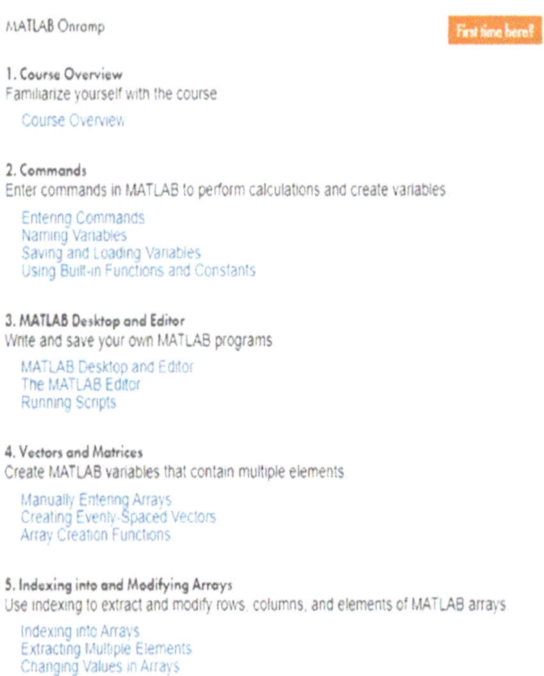

Figure 1.3: Screenshot of the "Onramp" MATLAB introduction.

1.3 Symbolic Math Tools, Introduction

Then one can run the Live script "Symbolic_X_LV", appended, for an introduction to MATLAB symbolic logic — tools which are heavily used in the text. The aim here is to use the symbolic math tools to avoid much of the algebraic equation work and rather concentrate on the physics.

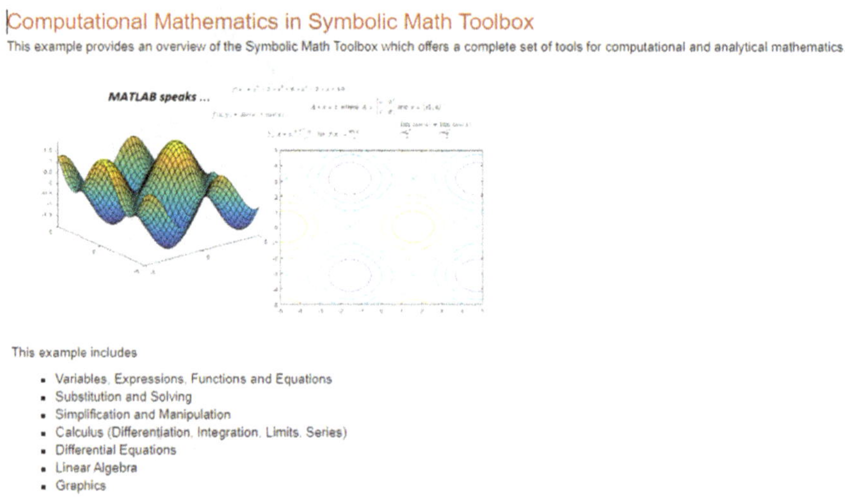

Figure 1.4: Screenshot of the "Symbolic_X_LV" output.

It is easiest to separately run the live script "Symbolic_X_LV" in order to see an introduction to MATLAB symbolic logic such as syms (defining symbolic variables), solve (solve algebraic equations), subs(substitute), simplify(simplify a symbolic expression), factor (factorize), diff (differentiate), int(integrate) and dsolve (solve a differential equation). These and other symbolic tools will be used heavily in the text, so that a quick introduction is a good first step.

```
% Symbolic_X_LV
```

1.4 Script Format, Editor

The opened scripts are displayed in the Live Editor. Apps have a distinct Editor. The formatting used by the editors is very useful for understanding the flow of computation. Data can be input to the Command Window by keyboard with prompt ≫. Text strings are in purple. The formatting is that comments are, %, and green. Symbolic variables and text characters are in purple and operations are executed in black. Flow control is in blue. Printing to the Command Line also appears inline for "Live" scripts which allows one to view the output on the same location as the script. This feature of "Live" scripts is very useful in following the flow of the calculation and it will typically be used in this text. Live Editor comments on warnings and errors appear on the right. If a statement is not followed by a %, the printout is inline with the Live script. The Live Editor "INSERT" tab allows added code, Control (Edit Fields, DropDown menus, and Numeric Sliders) to be inserted. Text and Equations can also be added to the Live code.

```
global xx ;   % global (common) variables are blue (unfavored now
in Matlab, use function arguments to pass variables)
%
for i = 1:2     % controls are in blue - indented by loops
    yy(i) = i *sin(i);
end
%
% any syntax errors appear in red, with script explanation on hover
% comments are in green
%
xx = linspace(0,2 *pi,100); % executables are black
%
```

A short answer to a query about a MATLAB function can be found using the help query in the Command Window or Live script. For example, help on the sin function. For more extensive documentation use the Search Documentation window in the Home tab. MATLAB has an extensive suite of special functions made available. Almost any mathematical function used in physics can be found in MATLAB and used as a library.

```
help('sin')
```

```
 sin    Sine of argument in radians.
   sin(X) is the sine of the elements of X.

    See also asin, sind, sinpi.

    Documentation for sin
    Other uses of sin
```

1.5 Symbolic Math Tools, Simple Examples

Some useful symbolic tools are shown below. The full suite of tools can be found using the "Search Documentation" window. ans is the answer if a ; does not suppress printing..

Differentiate

```
syms x
diff(sin(x)*cos(x)^2)
```

ans = $\cos(x)^3 - 2\cos(x)\sin(x)^2$

Integrate

```
syms x
int(x*cos(x)^2)
```

ans = $\frac{x\sin(2x)}{4} - \frac{\sin(x)^2}{4} + \frac{x^2}{4}$

Taylor expansion

```
taylor(exp(-x))
```

ans = $-\frac{x^5}{120} + \frac{x^4}{24} - \frac{x^3}{6} + \frac{x^2}{2} - x + 1$

Solve an algebraic equation or system of equations

```
syms a b c x y
solve( a*x^2+b*x+c == 0)
```

ans = $\begin{pmatrix} -\frac{b+\sqrt{b^2-4ac}}{2a} \\ -\frac{b-\sqrt{b^2-4ac}}{2a} \end{pmatrix}$

Series Summation

A series of terms can be easily summed. An example follows. A sum with a finite number of terms is also an option.

```
symsum(12/x^2,1,Inf)
```

ans = $2\pi^2$

Simplify, Pretty
Sometimes complex symbolic solutions need to be cleaned up. "simplify" and "pretty" are often useful. Another useful symbolic utility is "factor"

```
y = sin(x)^2+cos(x)^2 ;
simplify(y)
```

ans = 1

```
pretty(y)
```

$\cos(x)^2 + \sin(x)^2$

Factor a number

```
y = 300;
factor(y)
```

ans = 1 × 5

2	2	3	5	5

Matrix inversion can be used for both numeric and symbolic matrices

```
x = [1 0 2; -1 5 0; 0 3 -9];
inv(x)
```

ans = 3 × 3

0.8824	-0.1176	0.1961
0.1765	0.1765	0.0392
0.0588	0.0588	-0.0980

The eigenvectors, V, and eigenvalues, D, of numeric or symbolic matrices can be found using the utility "eig"

```
syms c
A = [c 1 0 0; 0 c 0 0; 0 0 3*c 0; 0 0 0 3*c];
[V,D] = eig(A)
```

$$V = \begin{pmatrix} 1 & 0 & 0 \\ 0 & 0 & 0 \\ 0 & 1 & 0 \\ 0 & 0 & 1 \end{pmatrix}$$

D = $\begin{pmatrix} c & 0 & 0 & 0 \\ 0 & c & 0 & 0 \\ 0 & 0 & 3c & 0 \\ 0 & 0 & 0 & 3c \end{pmatrix}$

ODE - Ordinary Differential Equation, or system of equations

```
syms y(t)
ode = diff(y,t)==t*y;
ysol(t) = dsolve(ode)
```

ysol(t) = $C_1 e^{\frac{t^2}{2}}$

This is a first order equation with no initial conditions specified, which means there is one unspecified constant, C1. If a symbolic solution does not exist the differential equation may be solved using a numerical solution employing "ode45". Many examples of these solutions appear throughout the text.

Fourier Transform

```
syms x t
fourier(exp(-x^2),x,t)
```

ans = $\sqrt{\pi} e^{-\frac{t^2}{4}}$

The Fourier transform of a Gaussian with r.m.s s is a Gaussian with r.m.s. 1/2*s. Other closed form solutions are sinx/x, exp(-|x|) and exp(-x)*heaviside(x).

1.6 ODE, Missile Tracking

As an example of the use of the ODE tool in MATLAB consider the problem of a missile that is on an intercept course with a target. The target is initially at $(x, y) = (L, 0)$ when sighted by the missile at $(0, 0)$. The missile is q times faster than the target so it will ultimately overtake it at some y and $x = 0$. The optimal strategy in this simplified case is always to aim at the present position of the target. The differential equation in this case is:

$$q(L - x)\frac{\mathrm{d}^2}{\mathrm{d}x^2}y = \sqrt{1 + (\mathrm{dy}/\mathrm{dx})^2} \tag{1.1}$$

This equation is solvable as shown below. The closed form solution for the intercept location at $x = L$ is:

$$y_{\text{int}} = L(q/(q^2 - 1)) \tag{1.2}$$

If q is very large the interception is immediate — at $y = 0$. If q is near one. the interception point is very far away. There is no solution if $q < 1$.

A movie of the missile and target paths is made. The distance L is fixed at 5.0. A "slider" is used to set the value of q. Limiting cases are $q = 1$, no interception, and $q \gg 1$, immediate interception at $y = 0$. In these scripts, many "comment" lines are given in order to explain what the code is doing so that the computational flow is clarified.

```
%
% Solve for missile seeking target of slower speed
%
close all ;                  % erase old plots
clear all;
%
syms y(x) vo t L q ys x
%
% Missile launched at (0,0) toward (L,0)
% Tracks by always aiming at present position of the target
% Missile velocity is q times the target velocity)
%
Dy = diff(y,x);
ode = diff(y,x,2)*q*(L-x) == sqrt(1+Dy^2);
cond1 = y(0) == 0;
cond2 = Dy(0) == 0;
conds = [cond1 cond2];
ySol(x) = dsolve(ode,conds);
ySol = simplify(ySol)
```

ySol(x) =

$$\frac{Lq}{q^2-1} - \frac{q\left(\cosh\left(\frac{\log(L)-\log(L-x)}{q}\right) + q\sinh\left(\frac{\log(L)-\log(L-x)}{q}\right)\right)(L-x)}{q^2-1}$$

```
LL = 5.0 ; % Location at y = 0 when target is acquired
qq = 1.3 ; %speed factor by which missile exceeds target
%
% intercept found analytically
yint = (LL .*qq) ./(qq .^2 - 1.0)
```

yint = 9.4203

```
%
xx = linspace(0,LL); % now make a movie
%
for i = 1:length(xx)
    x = xx(i);  % x of missile
    L = LL;
    q = qq;
    yy(i) = eval(ySol(x)); % y of missile
    xt(i) = LL;
    yt(i) = (i ./length(xx)) .*yint;
    %
end
%
figure
for i = 1:length(xx)
    plot(xx(i),yy(i),'bo',xt(i),yt(i),'r*')
    title('Movie of Missile and Target')
    axis([0 LL 0 yint])
    xlabel('x')
    ylabel('y')
    pause(0.1)
end
hold('on')
%
plot(xx,yy,'b-',xt,yt,'r:')
```

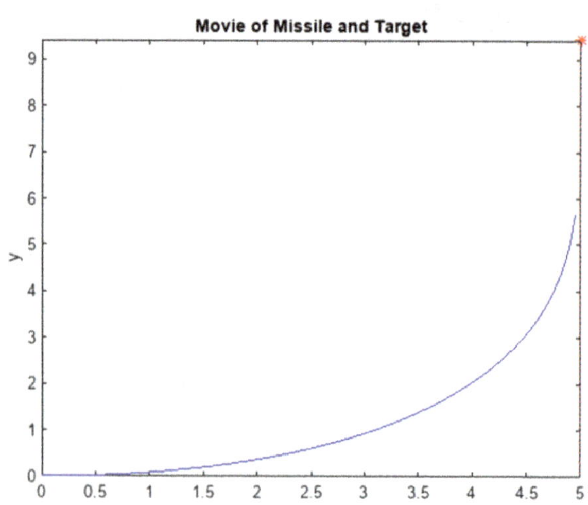

Figure 1.5: Movie and plot of the missile height as a function of the distance to the target. The movie can be replayed using button on the left.

```
%
qqq = linspace(1.1,5.0);
yint = (qqq) ./(qqq .^2 - 1.0);
figure
plot(qqq,yint,'b-')
title('Interception y as a Funtion of Speed Ratio')
xlabel('Missile/Target speed ratio')
ylabel('yint/L')
axis([1 max(qqq) 0 max(yint)])
```

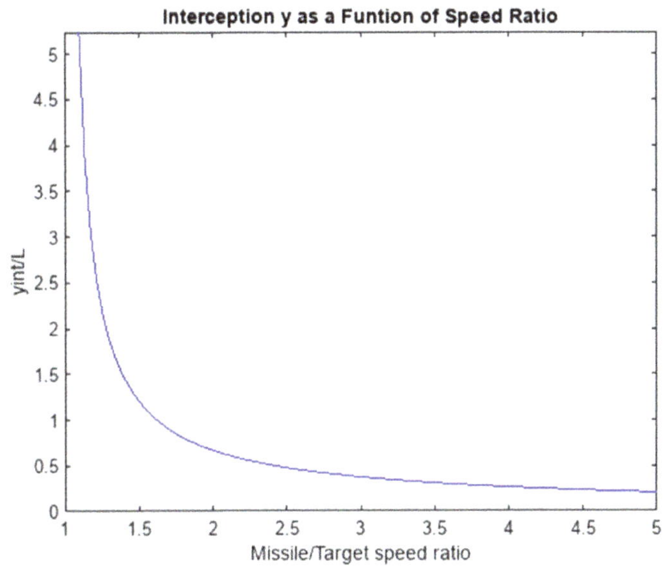

Figure 1.6: Interception distance as a function of q.

The meaning of the symbols used in this text are collected in Appendix 9.2. All the generated plots can be annotated after the fact. Appendix 9.3 shows the possibilities for annotation. The user has full freedom to explore and change any plot that has been generated.

1.7 MATLAB Plots

The "PLOTS" tab in the startup "Command Window" gives a list of possible plots in MATLAB. There are many options for 1, 2, and 3 variable plotting. A complete set is listed in the "Plots" tab. The first plots shown, in the section above, are generated using the "plot" tool for displaying the behavior of a single variable. Some options for 2 variable plotting are shown below. In addition active editing of Figure properties is available in the Command Window,

```
%
figure
colormap('jet')
%
% options mesh = 1, meshc = 2, surface = 3, surface contours = 3
Itype  = 4
```

```
Itype = 4
```

```
if Itype == 1
    [X,Y] = meshgrid(-3:.125:3);
    Z = peaks(X,Y);
    mesh(Z);
    title('Mesh')
end
%
if Itype == 2
    [X,Y] = meshgrid(-3:.125:3);
    Z = peaks(X,Y);
    meshc(Z);
    title('Mesh-Contour')
end
%
if Itype == 3
    [X,Y,Z] = peaks(30);
    surf(X,Y,Z);
    title('Surface')
end
%
if Itype  == 4
    [X,Y,Z] = peaks(30);
    surfc(X,Y,Z);
    title('Surface-Contour')
end
```

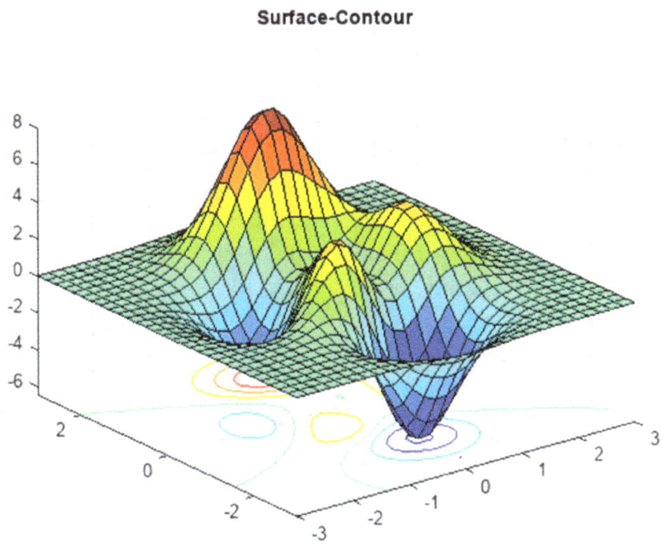

Figure 1.7: Example of a MATLAB 2 dimensional surface plot, with contours.

1.8 Monte Carlo, Analytic

Other symbolic math tools will be introduced when their use is dictated by the specific physics problem which is being addressed. Another useful tool is to make a model of a system by choosing distributions of variables used to represent the system. The MATLAB utility rand is a random number generator. In a few cases the distributions can be generated analytically. The script below uses a "drop down" menu to choose between five such functions and a "slider" is used to change a parameter which characterizes the distribution, which is then plotted.

For the five cases below an analytic inversion is possible. The function f with parameters α and the random number r can be solved for x in these special cases.

$$\int_{x_{\min}}^{x} f(x,\alpha)\mathrm{dx} \Big/ \int_{x_{\min}}^{x_{\max}} (f(x,\alpha))\mathrm{dx} = r \tag{1.3}$$

For example, one can solve for x as a function of r using the utility solve.

```
syms x xmin xmax a sol r
sol = solve(int(exp(-a*x),x,xmin)/int(exp(-a*x),xmin, xmax)-r ==0)
```

sol = $-\dfrac{\log\left(e^{-a\,\mathrm{xmin}} - r\left(e^{-a\,\mathrm{xmax}} - e^{-a\,\mathrm{xmin}}\right)\right)}{a}$

The other functions can also be inverted in the same fashion. The reader is encouraged to explore the other distributions.

```
% Monte Carlo - analytic method of inversion/integration
%
% Initialize
%close all
clear all
% Monte Carlo - Analytic Examples
% 'IType','Uniform','Power Law','Exp','Lorentz','Gaussian');
%
% analytic Monte Carlo generation of x values
% ai is the power x^a1, or exp(a1*x), or the Full width, or the
Gaussian r.m.s
%
xmi = 0.0 ;
xmx = 1.0 ;
a1 = 0.5;
val = "Lorentz";
figure
tf = strcmpi(val,'Uniform');
if tf == 1
    for i = 1:1000
      y(i) = rand;
    end
    histogram(y,50)
    title('1000 Uniformly Distributed (0,1) Points')
end
%
tf = strcmpi(val,'Power');
if tf == 1
     % a1 = Power Law Exponent
     b = a1 + 1;
     c = 1/b;
     % a cannot be -1, xmi cannot be 0, use 0.01
```

```
     for i = 1:1000
        y(i) = (xmi .^b + rand .*(xmx .^b - xmi .^b)) .^c;
     end
     histogram(y,50)
     title('1000 Power Law Points')
end
tf = strcmpi(val,'Exp') ;
if tf == 1
   %
   for i = 1:1000
       y(i) = -a1 .*log(exp(-xmi ./a1) + rand .*(exp(-xmx ./a1) -
       exp(-xmi ./a1)));
   end
   histogram(y,50)
   title('1000 Exponential Points')
end
tf = strcmpi(val,'Gauss') ;
if tf == 1
     a1 = 0.0; % r.m.s
     a2 = 0.5 % meam
     c = 2.0 .*a1 .*a1;  % pick r^2
     for i = 1:1000
          yy = sqrt(-2.0 .*a2 .*a2 .*log(rand));
          ph =  2 .*pi .*rand;
          y(i) = a1 + yy .*cos(ph);
     end
     histogram(y,50)
     title('1000 Gaussian Points')
end
tf = strcmpi(val,'Lorentz') ;
if tf == 1
    % a1 is mean, a2 is linewidth
    a1 = 0.5;
    a2 = 0.5;
    phmi = (2 .*(xmi - a1)) ./a2;
    phmx = (2 .*(xmx - a1)) ./a2;
    for i = 1:1000
        y(i) = a1 + (a2 ./2) .*tan(atan(phmi) + rand .*(atan(phmx)
        - atan(phmi)));
    end
    histogram(y,50)
    title('1000 Lorentzian Points')
end
```

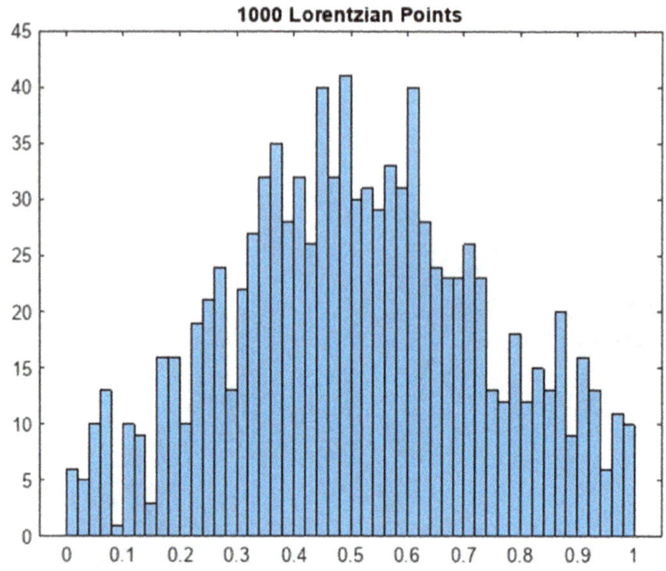

Figure 1.8: Histogram of the Monte Carlo Lorentzian distribution.

1.9 Monte Carlo, Numeric

Most distributions cannot be chosen by inversion. In that case one simple, but inefficient, method uses 2 random numbers. A random number r_1 is chosen over the chosen x range where f has a maximum value $f_{\max}$. A second random number r_2 is found and if $r_2 f_{\max} < f(x)$ that value of x is accepted. A specific example of the angular distribution for Compton scattering follows. The incoming photon energy is chosen by "slider". The reader is, as always encouraged to explored the shape of the distribution as a function of the photon energy, with a scale set by the electron target mass.

```
%
% Monte Carlo - numerical method of inversion/integration
%
%close all ;                  % erase old plots
clear all;
%
% Initialize
%
% Monte Carlo - Numerical Compton Scattering Example
% Photon Energy Divided by Electron Mass
w = 3.1;
```

```
%
ctmi = -1;  % Compton scattering angle
ctmx = 1;
pmx = 2;
pmi = 0;
ipass = 0;
for i = 1:10000
    ct = ctmi + rand .*(ctmx - ctmi);
    yc = 1.0 ./(1 + w .*(1-ct));
    pc = yc .*yc .*(yc + 1 ./yc - (1 - ct .*ct));
    % Compton dsigma/domega
    if rand < pc ./pmx;
        ipass = ipass + 1;
        Compt(ipass) = ct;
    end
end
figure
histogram(Compt,50);
title('Compton Scattering Angular Distribution')
xlabel('cos(\theta)')
ylabel('d\sigma/d\Omega')
```

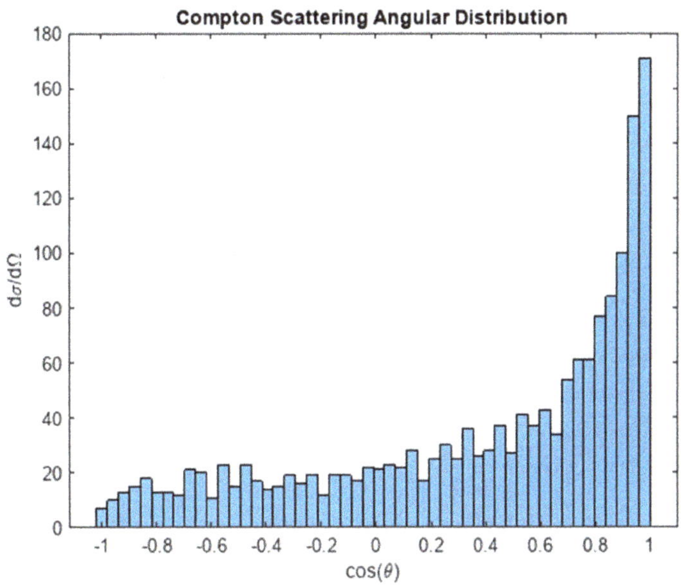

Figure 1.9: Histogram of the numeric generated Compton angular distribution.

1.10 Fourier Series

Any periodic function can be represented by a series of harmonic functions. The Fourier series uses cos and sin functions. The series terms are shown in Eq. (1.4).

$$x = a_o/2 + \sum_k [a_k \cos(k\omega t) + b_k \sin(k\omega t)]$$

$$u = t/T, \quad \omega t = 2\pi u, \quad a_k = 2\int x(u)\cos(2\pi \mathrm{ku})\mathrm{du},$$

$$b_k = 2\int x(u)\sin(2\pi \mathrm{ku})\mathrm{du}, \tag{1.4}$$

The expansion is over one period, τ, of the function $y(u)$. The constant coefficient, a_o, is just the average value of $y(u)$ over the period. The other coefficients are calculable, either symbolically or numerically. Examples are shown using the App "Fourier_Series_App" of a few functions which are chosen using the "DropDown" menu. A sawtooth wave function is shown in Figure 1.10, where the function, the series and the series coefficients are displayed in Figure 1.11. The movie with increasing number of terms shows how the series more closely approximates the function. The user is encouraged to look at the functions that are supplied and to see how well the function is represented as the number of terms in the series varies up to 6. The constant term, ao, is not displayed only the first five terms aj.

An arbitrary function can be made out of a series of sin and cos functions with different frequencies, $k\omega$ where k is an integer and with coefficients a_k and b_k. The function is periodic with period T, or $\omega = 2\pi/T$. The series is x. The coefficients are easily found using the MATLAB symbolic tool "int", The plots show how the series better approximates the function as the number terms increases. Series with terms up to 6 are evaluated and plotted. The functions supplied are a square wave, a triangular wave and a sawtooth.

```
%close all
%clear all
%
% Fourier series demo
%
% x = ao/2 + sumk( ak*cos(kwt) + bk*sin(kwt) )
% ak = 2*int(x(u)*cos(2*pi*k*u)du, bk = 2*int(x(u)*sin(2*pi*k*u)du
% u = t/T, [-1/2,1/2], w = 2*pi/T, w*t = 2*pi*u  \n')
```

```
%
% Type of Function','Square Wave','Triangle','Sawtooth');
%
u = linspace(-0.5,0.5,50); % one period
%
val = "Saw";
tf = strcmpi(val,'Square');
if tf == 1
    Itype = 1;
end
tf = strcmpi(val,'Triangle');
if tf == 1
    Itype = 2;
end
tf = strcmpi(val,'Saw');
if tf == 1
    Itype = 3;
end
%
% form the square wave
%
if Itype == 1
    for k = 1:length(u)
        x(k) = 0.0;
        if abs(u(k)) < 0.25;
            x(k) = 1.0;
        end
    end
    %
    % 5 terms + constant value
    %
    ao = 1;
    xf1 = ao ./2.0;
    for k = 1:5
        z(k) = (pi .*k) ./2.0;
        a(k) = sin(z(k)) ./(z(k));
        b(k) = 0.0;
    end
    % terms are z = pi*k/2, bk = 0, ak = sin(z)/z \n')
end
%
if Itype == 2
    %
    % form the triangle
    %
```

```
    for k = 1:length(u)
        if u(k) < 0.0
            x(k) = 1.0 + 2.0 .*u(k);
        else
            x(k) = 1.0 - 2.0 .*u(k);
        end
    end
    ao = 1;
    xf1 = ao ./2.0;
    for k = 1:5
        z(k) = (pi .*k);
        a(k) = 2.0 .*(1.0 - cos(z(k)))  ./(z(k) .^2);
        b(k) = 0.0;
    end
    % termsz = pi*k, bk = 0, ak = 2*(1-cosz)/z^2 \n')
end
%
if Itype == 3
    %
    % form the sawtooth
    %
    x = u;
    ao = 0;
    xf1 = ao ./2.0;
    for k = 1:5
        z(k) = (pi .*k);
        b(k) = (sin(z(k)) - z(k) .*cos(z(k))) ./(z(k) .^2);
        a(k) = 0.0;
    end
    % termsz = pi*k, bk = 0, ak = 2*(1-cosz)/z^2 \n')
end
%
% the number of erms can be changes by changing the i loop limit
from the
% default value of 6
%
for i = 2:6
    for j = 1:length(u)
        if i == 2
            xf(i,j) = xf1 + a(i-1) .*cos(2 .*pi .*u(j)) + b(i-1)
            .*sin(2 .*pi .*u(j));
        else
            xf(i,j) = xf(i-1,j) + a(i-1) .*cos(2 .*(i-1) .*pi
            .*u(j)) + b(i-1) .*sin(2 .* (i-1) .*pi .*u(j));
        end
```

```
    end
end
figure
for i = 2:6
    plot(u,x,'-b',u,xf(i,:),'-r')
    title('Function and Series')
    xlabel('u = t/T')
    ylabel('x and Series for 2 terms')
    hold('on')
    %
    pause(3)
end
hold('off')
```

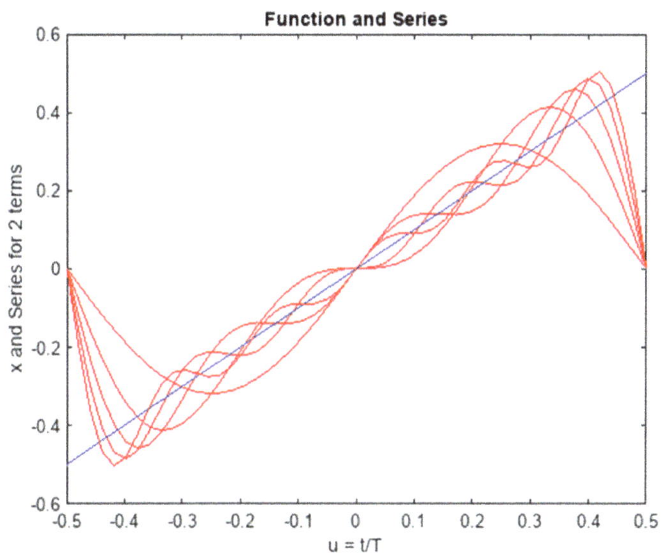

Figure 1.10: Overlay plot of the Fourier series fit to a “sawtooth”.

```
%
figure
iterm = [1 2 3 4 5 ];
plot(iterm,a,'bo', iterm,b,'r*')
title('Series Coefficients')
xlabel('# of terms')
ylabel('cos and sin terms')
legend('cos','sin')
```

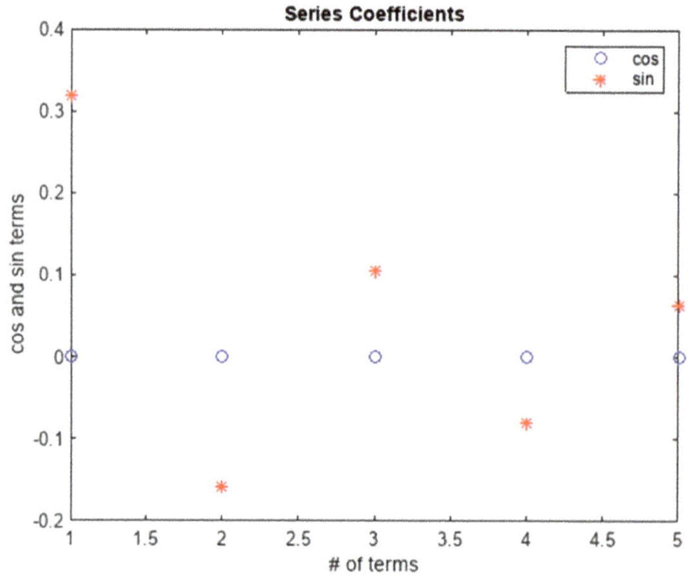

Figure 1.11: Cos and sin coefficients for the fit shown in Fig. 1.10 above.

The reader can explore the other shapes using the Edit field. The number of terms could also be changed by editing the code.

1.11 Polynomial Data Fits

In many applications it is important to "fit" to data with some hypothesis or simply to have a convenient way to represent the data. MATLAB supplies a polynomial fit to the data. Note that the data points have no errors and therefore the fits are not weighted by the errors on the individual data points. The simplest fit is a constant and the power of the polynomial can be easily varied.

```
%
% Look at polynomial fits to fake data
%
%close all
clear all;
%
% Data Fit Tools in MATLAB, Polynomial, polyfit, polyval
% data on temp along rod between 2 heat baths at fixed temperatures
```

```
%
x = 1:9;
T = [ 15.6 17.5 36.6 43.8 58.2 61.6 64.2 70.4 98.8];
for i = 1:9
    sig(i) = 6.7; % no error weighting in polyfit
end
%
figure
errorbar(x,T,sig,'o')
title('Temperature Along a Bar Between 2 Heat Baths')
xlabel('x(cm)')
ylabel('T(^oC)')
%
% Polynomial Coefficients, choose order
%
npol = 3;
P1 = polyfit(x,T,npol);
Tf = polyval(P1,x);
hold('on')
plot(x,Tf,'r')
hold('off')
```

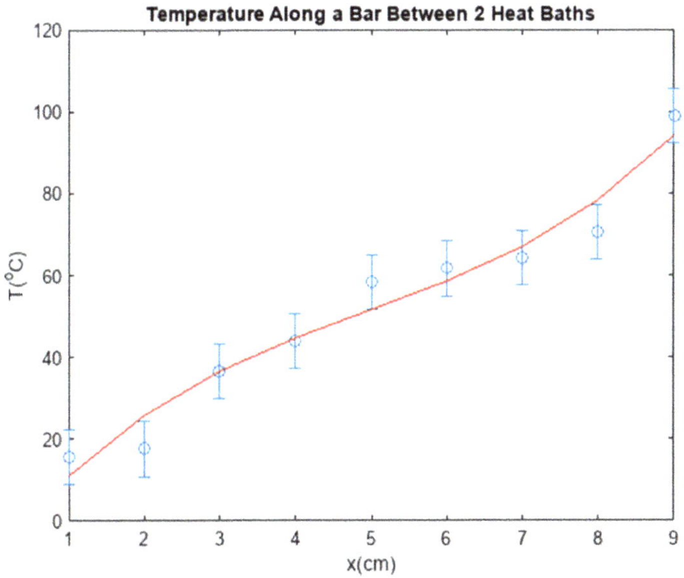

Figure 1.12: Fit to data with a polynomial.

More complex data fitting is shown in Appendix 9.4, where a general non-linear chisquared fitting package is provided. This package is quite general, and the user is free to define both the data set, with errors, to be fit and the function to be fit to.

1.12 Euler Angles, Simplified

For a solid body there are 3 angles that relate the axes of the body to the axis of a coordinate system. These angles will be described and used in a later discussion of the motion of a top. For now, the 2 angles relating a vector $\overrightarrow{p}$ to the vector $\overrightarrow{pp}$ in a frame defined by the spherical polar coordinates θ, ϕ are a special simple case. For example these transformations can be used to relate the "lab frame" to the frame where a particle moving along a z axis decays into a particle with vector momentum $\overrightarrow{p}$ which is useful in transforming coordinate frames for particles that scatter or decay.

$$\begin{bmatrix} \mathrm{pp}_x \\ \mathrm{pp}_y \\ \mathrm{pp}_z \end{bmatrix} = \begin{bmatrix} \cos\theta\cos\phi & -\sin\phi & \sin\theta\cos\phi \\ \cos\theta\sin\phi & \cos\phi & \sin\theta\sin\phi \\ \sin\theta & 0 & \cos\theta \end{bmatrix} \begin{bmatrix} p_x \\ p_y \\ p_z \end{bmatrix} \tag{1.5}$$

1.13 Reading an Equation

In the next sections many different physics topics are covered. Their characteristics are governed by specific equations. It is important to first look at the equation before trying to solve it. For example one can examine the dimensions of the constants and variables that appear. If a solution is not dimensionally correct it must be wrong. For example explore the Schrodinger equation, which appears in quantum mechanics.

$$\left[-\hbar^2\left(\partial^2\psi/\partial^2 x\right)/2m\right] + V\psi = i\hbar\partial\psi/\partial t \tag{1.6}$$

The potential energy V term is set by the dynamics. For example it is known from atomic spectroscopy to be $\langle V\rangle \sim 1\,\mathrm{eV}$. That then tells us that $\hbar/t$ has the dimensions of energy, so the dimension of $\hbar$ must be energy times time. Then since energy times time has the dimension of mass times length squared divided by time, the dimensions of $\hbar$ are those of position times momentum. That is known from the uncertainty relationship for energy and time and that for position and momentum Then one can see what the characteristic lengths and times of objects that obey this equation are with $\langle V \rangle \sim 1\,\text{eV}$. The time scale is $\langle t \rangle \sim \hbar/\langle V \rangle = 6.6 \times 10^\wedge{-16}$ s or ~1 fs. Regarding length, $\hbar c = 200\,\text{eV}\,\text{nm}$, and $m_e c^2 = 5.11 \times 10^5\,\text{eV}$. A typical physical size for the electrons in an atomic system is $\langle x^2 \rangle \sim (\hbar c)^2/m_e c^2 \langle V \rangle \sim 0.08\,\text{nm}$, which is indeed the typical size of an atom. So, already we have an idea of the magnitudes of the answers that arise from actually solving the equation.

First, one can attempt to find a symbolic solution. Only if that fails does one need a full numerical solution. (MATLAB has such, called ode45). Before solving the full equation, often a solution can be found by making a simplifying assumption. For example a free particle, $V = 0$, has plane wave solutions with wave number and frequency related by $(\hbar k)^2 = 2m\hbar\omega$.

1.14 MATLAB Supplied "Cheat-Sheet"

MATLAB has compiled a very useful summary of the utilities that are often used. This "cheat-sheet" is appended here as a quick look up for some of the more important utilities that are available. For more detail, the command line "help" or the "Search Documentation" window are available in the Command Window. The reader can now, if wished, skip to physics in Chapter 2 and consult the "Cheatsheet" only when specifically necessary.

MATLAB Environment	
clc	Clear command window
help fun	Display in-line help for **fun**
doc fun	Open documentation for **fun**
load("filename","vars")	Load variables from **.mat** file
uiimport("filename")	Open interactive import tool
save("filename","vars")	Save variables to file
clear item	Remove items from workspace
examplescript	Run the script file named **examplescript**
format style	Set output display format
ver	Get list of installed toolboxes
tic, toc	Start and stop timer
Ctrl+C	Abort the current calculation

Defining and Changing Array Variables	
`a = 5`	Define variable a with value 5

`A = [1 2 3; 4 5 6]`
`A = [1 2 3`
`     4 5 6]`
	Define A as a 2x3 matrix
	"space" separates columns
	";" or new line separates rows

`[A,B]`	Concatenate arrays horizontally
`[A;B]`	Concatenate arrays vertically
`x(4) = 7`	Change 4th element of x to 7
`A(1,3) = 5`	Change A(1,3) to 5
`x(5:10)`	Get 5th to 10th elements of x
`x(1:2:end)`	Get every 2nd element of x (1st to last)
`x(x>6)`	List elements greater than 6
`x(x==10)=1`	Change elements using condition
`A(4,:)`	Get 4th row of A
`A(:,3)`	Get 3rd column of A
`A(6, 2:5)`	Get 2nd to 5th element in 6th row of A
`A(:,[1 7])=A(:,[7 1])`	Swap the 1st and 7th column
`a:b`	[a, a+1, a+2, ..., a+n] with a+n≤b
`a:ds:b`	Create regularly spaced vector with spacing ds
`linspace(a,b,n)`	Create vector of n equally spaced values
`logspace(a,b,n)`	Create vector of n logarithmically spaced values
`zeros(m,n)`	Create m x n matrix of zeros
`ones(m,n)`	Create m x n matrix of ones
`eye(n)`	Create a n x n identity matrix
`A=diag(x)`	Create diagonal matrix from vector
`x=diag(A)`	Get diagonal elements of matrix
`meshgrid(x,y)`	Create 2D and 3D grids
`rand(m,n), randi`	Create uniformly distributed random numbers or integers
`randn(m,n)`	Create normally distributed random numbers

Operators and Special Characters	
`+, -, *, /`	Matrix math operations
`.*, ./`	Array multiplication and division (element-wise operations)
`^, .^`	Matrix and array power
`\`	Left division or linear optimization
`.', '`	Normal and complex conjugate transpose
`==, ~=, <, >, <=, >=`	Relational operators
`&&, ||, ~, xor`	Logical operations (AND, NOT, OR, XOR)
`;`	Suppress output display
`...`	Connect lines (with break)
`% Description`	Comment
`'Hello'`	Definition of a character vector
`"This is a string"`	Definition of a string
`str1 + str2`	Append strings

Special Variables and Constants	
`ans`	Most recent answer
`Pi`	π=3.141592654...
`i, j, 1i, 1j`	Imaginary unit
`NaN, nan`	Not a number (i.e., division by zero)
`Inf, inf`	Infinity
`eps`	Floating-point relative accuracy

Complex Numbers	
`i, j, 1i, 1j`	Imaginary unit
`real(z)`	Real part of complex number
`imag(z)`	Imaginary part of complex number
`angle(z)`	Phase angle in radians
`conj(z)`	Element-wise complex conjugate
`isreal(z)`	Determine whether array is real

Elementary Functions	
`sin(x), asin`	Sine and inverse (argument in radians)
`sind(x), asind`	Sine and inverse (argument in degrees)
`sinh(x), asinh`	Hyperbolic sine and inverse (arg. in radians)
Analogous for the other trigonometric functions: `cos`, `tan`, `csc`, `sec`, and `cot`	
`abs(x)`	Absolute value of x, complex magnitude
`exp(x)`	Exponential of x
`sqrt(x), nthroot(x,n)`	Square root, real nth root of real numbers
`log(x)`	Natural logarithm of x
`log2(x), log10`	Logarithm with base 2 and 10, respectively
`factorial(n)`	Factorial of n
`sign(x)`	Sign of x
`mod(x,d)`	Remainder after division (modulo)
`ceil(x), fix, floor`	Round toward +inf, 0, -inf
`round(x)`	Round to nearest decimal or integer

Plotting	
plot(x,y,LineSpec) Line styles: **-, --, :, -.** Markers: **+, o, *, ., x, s, d** Colors: **r, g, b, c, m, y, k, w**	Plot y vs. x (**LineSpec** is optional) **LineSpec** is a combination of **linestyle, marker**, and **color** as a string. Example: **"-r"** = red solid line without markers
title("Title")	Add plot title
legend("1st", "2nd")	Add legend to axes
x/y/zlabel("label")	Add x/y/z axis label
x/y/zticks(ticksvec)	Get or set x/y/z axis ticks
x/y/zticklabels(labels)	Get or set x/y/z axis tick labels
x/y/ztickangle(angle)	Rotate x/y/z axis tick labels
x/y/zlim	Get or set x/y/z axis range
axis(lim), axis style	Set axis limits and style
text(x,y,"txt")	Add text
grid on/off	Show axis grid
hold on/off	Retain the current plot when adding new plots
subplot(m,n,p), **tiledlayout(m,n)**	Create axes in tiled positions
yyaxis left/right	Create second y-axis
figure	Create figure window
gcf, gca	Get current figure, get current axis
clf	Clear current figure
close *all*	Close open figures

Tables	
`table(var1,...,varN)`	Create table from data in variables var1, ..., varN
`readtable("file")`	Create table from file
`array2table(A)`	Convert numeric array to table
`T.var`	Extract data from variable var
`T(rows,columns)`, `T(rows,["col1","coln"])`	Create a new table with specified rows and columns from T
`T.varname=data`	Assign data to (new) column in T
`T.Properties`	Access properties of T
`categorical(A)`	Create a categorical array
`summary(T), groupsummary`	Print summary of table
`join(T1, T2)`	Join tables with common variables

Tasks (Live Editor)

Live Editor tasks are apps that can be added to a live script to interactively perform a specific set of operations. Tasks represent a series of MATLAB commands. To see the commands that the task runs, show the generated code.

Common tasks available from the Live Editor tab on the desktop toolstrip:

- Clean Missing Data
- Find Change Points
- Remove Trends
- Clean Outlier
- Find Local Extrema
- Smooth Data

Common Plot Types

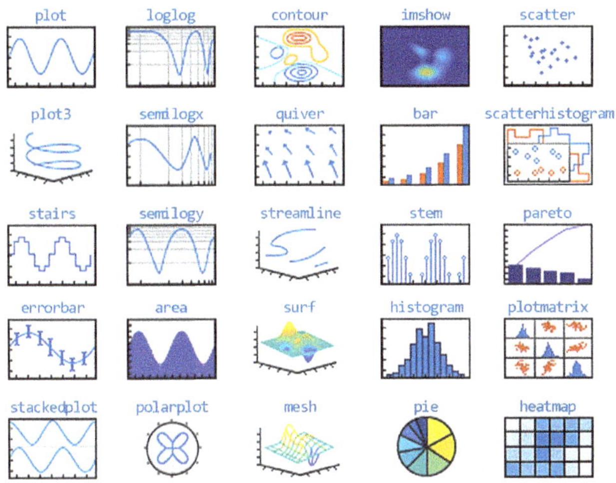

Plot Gallery: *mathworks.com/products/matlab/plot-gallery*

Programming Methods

Functions

```
% Save your function in a function file or at the end
% of a script file. Function files must have the
% same name as the 1st function
function cavg = cumavg(x) %multiple args. possible
    cavg=cumsum(vec)./(1:length(vec));
end
```

Anonymous Functions

```
% defined via function handles
fun = @(x) cos(x.^2)./abs(3*x);
```

Control Structures

if, elseif Conditions

```
if n<10
    disp("n smaller 10")
elseif n<=20
    disp("n between 10 and 20")
else
    disp("n larger than 20")
```

Switch Case

```
n = input("Enter an integer: ");
switch n
    case -1
        disp("negative one")
    case {0,1,2,3} % check four cases together
        disp("integer between 0 and 3")
    otherwise
        disp("integer value outside interval [-1,3]")
end % control structures terminate with end
```

For-Loop

```
% loop a specific number of times, and keep
% track of each iteration with an incrementing
% index variable
for i = 1:3
    disp("cool");
end % control structures terminate with end
```

While-Loop

```
% loops as long as a condition remains true
n = 1;
nFactorial = 1;
while nFactorial < 1e100
    n = n + 1;
    nFactorial = nFactorial * n;
end % control structures terminate with end
```

Further programming/control commands

`break`	Terminate execution of for- or while-loop
`continue`	Pass control to the next iteration of a loop
`try, catch`	Execute statements and catch errors

Numerical Methods	
`fzero(fun,x0)`	Root of nonlinear function
`fminsearch(fun,x0)`	Find minimum of function
`fminbnd(fun,x1,x2)`	Find minimum of fun in [x1, x2]
`fft(x), ifft(x)`	Fast Fourier transform and its inverse

Integration and Differentiation	
`integral(f,a,b)`	Numerical integration (analogous functions for 2D and 3D)
`trapz(x,y)`	Trapezoidal numerical integration
`diff(X)`	Differences and approximate derivatives
`gradient(X)`	Numerical gradient
`curl(X,Y,Z,U,V,W)`	Curl and angular velocity
`divergence(X,..,W)`	Compute divergence of vector field
`ode45(ode,tspan,y0)`	Solve system of nonstiff ODEs
`ode15s(ode,tspan,y0)`	Solve system of stiff ODEs
`deval(sol,x)`	Evaluate solution of differential equation
`pdepe(m,pde,ic,...` `bc,xm,ts)`	Solve 1D partial differential equation
`pdeval(m,xmesh,...` `usol,xq)`	Interpolate numeric PDE solution

Interpolation and Polynomials	
`interp1(x,v,xq)`	1D interpolation (analogous for 2D and 3D)
`pchip(x,v,xq)`	Piecewise cubic Hermite polynomial interpolation
`spline(x,v,xq)`	Cubic spline data interpolation
`ppval(pp,xq)`	Evaluate piecewise polynomial
`mkpp(breaks,coeffs)`	Make piecewise polynomial
`unmkpp(pp)`	Extract piecewise polynomial details
`poly(x)`	Polynomial with specified roots x
`polyeig(A0,A1,...,Ap)`	Eigenvalues for polynomial eigenvalue problem
`polyfit(x,y,d)`	Polynomial curve fitting
`residue(b,a)`	Partial fraction expansion/decomposition
`roots(p)`	Polynomial roots
`polyval(p,x)`	Evaluate poly p at points x
`polyint(p,k)`	Polynomial integration
`polyder(p)`	Polynomial differentiation

Matrices and Arrays	
`length(A)`	Length of largest array dimension
`size(A)`	Array dimensions
`numel(A)`	Number of elements in array
`sort(A)`	Sort array elements
`sortrows(A)`	Sort rows of array or table
`flip(A)`	Flip order of elements in array
`squeeze(A)`	Remove dimensions of length 1
`reshape(A,sz)`	Reshape array
`repmat(A,n)`	Repeat copies of array
`any(A), all`	Check if any/all elements are nonzero
`nnz(A)`	Number of nonzero array elements
`find(A)`	Indices and values of nonzero elements

Linear Algebra	
`rank(A)`	Rank of matrix
`trace(A)`	Sum of diagonal elements of matrix
`det(A)`	Determinant of matrix
`poly(A)`	Characteristic polynomial of matrix
`eig(A), eigs`	Eigenvalues and vectors of matrix (subset)
`inv(A), pinv`	Inverse and pseudo inverse of matrix
`norm(x)`	Norm of vector or matrix
`expm(A), logm`	Matrix exponential and logarithm
`cross(A,B)`	Cross product
`dot(A,B)`	Dot product
`kron(A,B)`	Kronecker tensor product
`null(A)`	Null space of matrix
`orth(A)`	Orthonormal basis for matrix range
`tril(A), triu`	Lower and upper triangular part of matrix
`linsolve(A,B)`	Solve linear system of the form AX=B
`lsqminnorm(A,B)`	Least-squares solution to linear equation
`qr(A), lu, chol`	Matrix decompositions
`svd(A)`	Singular value decomposition
`gsvd(A,B)`	Generalized SVD
`rref(A)`	Reduced row echelon form of matrix

Descriptive Statistics	
`sum(A), prod`	Sum or product (along columns)
`max(A), min, bounds`	Largest and smallest element
`mean(A), median, mode`	Statistical operations
`std(A), var`	Standard deviation and variance
`movsum(A,n), movprod, movmax, movmin, movmean, movmedian, movstd, movvar`	Moving statistical functions n = length of moving window
`cumsum(A), cumprod, cummax, cummin`	Cumulative statistical functions
`smoothdata(A)`	Smooth noisy data
`histcounts(X)`	Calculate histogram bin counts
`corrcoef(A), cov`	Correlation coefficients, covariance
`xcorr(x,y), xcov`	Cross-correlation, cross-covariance
`normalize(A)`	Normalize data
`detrend(x)`	Remove polynomial trend
`isoutlier(A)`	Find outliers in data

Symbolic Math*	
`sym x, syms x y z`	Declare symbolic variable
`eqn = y == 2*a + b`	Define a symbolic equation
`solve(eqns,vars)`	Solve symbolic expression for variable
`subs(expr,var,val)`	Substitute variable in expression
`expand(expr)`	Expand symbolic expression
`factor(expr)`	Factorize symbolic expression
`simplify(expr)`	Simplify symbolic expression
`assume(var,assumption)`	Make assumption for variable
`assumptions(z)`	Show assumptions for symbolic object
`fplot(expr), fcontour, fsurf, fmesh, fimplicit`	Plotting functions for symbolic expressions
`diff(expr,var,n)`	Differentiate symbolic expression
`dsolve(deqn,cond)`	Solve differential equation symbolically
`int(expr,var,[a, b])`	Integrate symbolic expression
`taylor(fun,var,z0)`	Taylor expansion of function

Chapter 2

Mechanics

"The most exciting phrase to hear in science, the one that heralds new discoveries, is not 'Eureka!' but 'That's funny'..."
Isaac Asimov

"Measure what can be measured, and make measureable what cannot be measured." Galileo Galilei

The first physics section concerns mechanics. This topic is, perhaps, most familiar to the user. The basic force for gravitation was discovered by Newton. In the context of planetary orbits, the equation of motion will be "read" in section 2.6. The initial topics in mechanics explore harmonic motion and pendula. Galileo in fact formulated a law for the period of a pendulum while in church observing the periods of candelabra and using his pulse rate to measure the periods.

Including all the code from the scripts makes the text too long. In this and subsequent Sections much of the code will be excised. It is fully available in the scripts which are an intrinsic part of the full text. The reader will, however, not be able to change the parameters of a problem, nor see the details of how the solutions are obtained. This reduction in text is unavoidable but in no way limits the reader from exploring each problem with a specific Live file supplied with the text.

2.1 Harmonic Oscillator — Free, Damped, Driven

Mechanics will be concerned largely with the gravitational force and the conservation laws for systems which have no dissipation so that energy and momentum are conserved. The force between 2 masses can be derived from a potential energy, $F_{12} = \mathrm{GM}_1 M_2/r_{12}$, $V_{12} = F_{12} r_{12}$. A potential energy felt by particle 1 often has the m_1 factored out, $\Phi = V/m_1$. For systems of particles there are a conserved energy, potential plus kinetic, $T = \mathrm{mv}^2/2$, and a conserved momentum, $\overrightarrow{p} = \mathrm{m}\overrightarrow{\mathrm{v}}$.

Start with the traditional simple harmonic oscillator (SHO) with damped harmonic motion but no driving force. In the undamped case a dimensional analysis (reading the equation) indicates that k/m has the dimensions of inverse time squared. Indeed, the undamped frequency is $\omega_o = \sqrt{k/m}$.

$$\frac{\mathrm{d}^2}{\mathrm{d}t^2}x = -\mathrm{kx}/m + (b/m)\frac{\mathrm{d}}{\mathrm{d}t}x \tag{2.1}$$

The equation is solved symbolically using the MATLAB utility "dsolve". With damping the solution can be underdamped or overdamped. The response frequencies are $\omega_d = \sqrt{\omega_o^2 - (b/2m)^2}$ leading to oscillatory or exponential solutions in the underdamped and overdamped cases. The "movie" of the motion of x can be played back to watch the solution evolve from a damped oscillation to a monotonic falloff as the damping term increases. A slow playback speed shows the full transition nicely and the playback can be halted at any interesting point.

```
%
% SHO with damping - MATLAB symbolic diff Eq, dsolve
% Damped Harmonic Oscillator
% d2x/dt2 = kx/m -b*dxdt/m, wo^2 = k/m, frictional damping, depends
in dx/dt
%
syms x(t) Dx D2x k m A b
Dx = diff(x,t);
D2x = diff(x,t,2);
ode = D2x == -(k*x)/m
```

ode(t) = $\frac{\partial^2}{\partial t^2}x(t) = -\frac{kx(t)}{m}$

```
cond1 = x(0) == A; cond2 = Dx(0) == 0;
conds = [cond1 cond2];
%
% Undamped Harmonic Oscillator, x(o) = A, dx(0) = 0
xx = dsolve(ode, conds );
xs = simplify(xx)
```

xs = $A\cos\left(\frac{\sqrt{k}t}{\sqrt{m}}\right)$

```
%
% Damped Harmonic Oscillator
ode = D2x == -(k*x)/m + b*Dx/m
```

ode(t) = $\frac{\partial^2}{\partial t^2}x(t) = \frac{b\frac{\partial}{\partial t}x(t)}{m} - \frac{kx(t)}{m}$

```
cond1 = x(0) == A; cond2 = Dx(0) == 0;
conds = [cond1 cond2];
xxd = dsolve(ode, conds );
x = simplify(xxd)
```

x = $\frac{Ae^{\frac{t(b-\sigma_1)}{2m}}(b+\sigma_1)}{2\sigma_1} - \frac{Ae^{\frac{t(b+\sigma_1)}{2m}}(b-\sigma_1)}{2\sigma_1}$

where

$\sigma_1 = \sqrt{b^2 - 4km}$

```
%
% d = b/(2*m), D = d/wo, x(t) = A*exp(-d*t)*cos(wt), w^2 = wo^2 -
d^2, underdamped
% x(t) = A*exp[-d +- sqrt(d^2 - wo^2)]*t, overdamped
% Example: R,L, C circuit, wo^2 = 1/(LC), d = R/(2L)
%
```

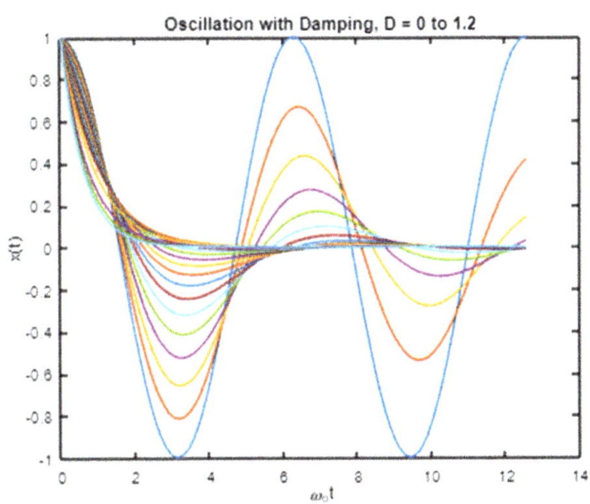

Figure 2.1: Plot of the amplitude of a damped oscillator for different damping coefficients.

A driven oscillator may be forced into resonance if the frequency of the driving term, ω_e, is the natural frequency. The script scans an oscillator with different damping factors D, driven by a term with driving frequency ω_e/ω_o from 0 to 2. The damping term is also varied and the response is plotted as ω_e is scanned. The maximum amplitude of the resonance and the width of the response depends on the damping. Lightly damped oscillators respond very strongly to the driving force near the natural frequency, while heavily damped oscillators respond less but over a wider range of frequencies. $y = 1/\sqrt{a+b}$, $a = [(\omega_e/\omega_o)^2 - 1]^2$, $b = (2D\omega_e/\omega_o)^2$. The maximum amplitude, y, occurs when $\omega_e = \omega_o$ and is formally infinite if the damping is zero.

```
% use the animation playback for the movies as available
% driven damped oscillator, driving amplitude = B, frequency = we
% scan over driving frequency and adjust damping factor, units with
m = 1
% Damped, Driven Oscillator - Resonance Behavior, Steady State
Solution
```

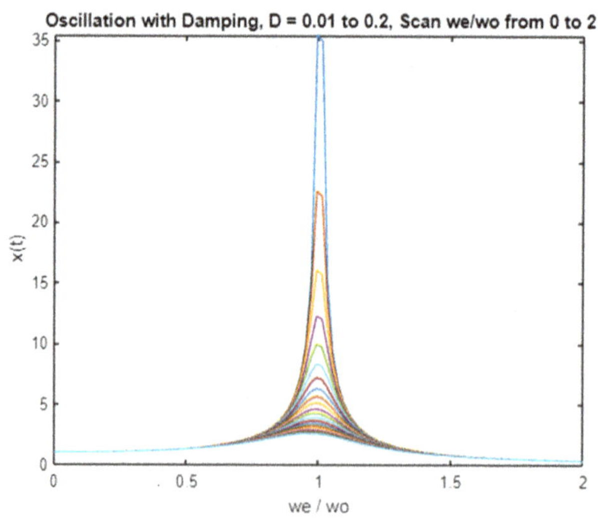

Figure 2.2: Plot of the oscillator displacement for a driven oscillator for different damping coefficients.

2.2 Two Coupled Pendula

Look at 2 coupled pendula. One has a fixed attachment point while the second is suspended from the heavy fob of the first. The 2 lengths and weights are taken to be equal to 1 for simplicity. Use the numerical solver "ode45" for their motion. Explore extreme motion with large initial angles. See if chaotic motion can be induced. For small angles the motion is simple; oscillations with the 2 pendula fully out of phase For larger angles the motion does not fully repeat over the times explored. The actual equations of motion are shown in th e ode45 function "pend2" but are not particularly illuminating in themselves. The user chooses the 2 initial angles of the 2 pendula. The initial velocities are set to 0. A "movie" is made of the motion of the 2 pendula. It is clear that quite nonlinear behavior can be created by choosing large initial angles even though the initial velocities are set to zero. The motion is constrained so that the fobs cannot be driven to large distances, but the second fob can wind itself up by rotating above the first fob.

```
%
% motion of 2 coupled pendula
% "chaotic" large angle motion
% Set initial position of pendula
% Initial Angles (degrees), Velocities = 0,[th1, th2]: ');
th1 = 80;
th2 = -90;
tho = [th1 th2];
tho = (tho .*pi) ./180.0;    % Convert angle to radians
% L in MKS units, m = 1 and g/L = 1
L = 1;
vo = [0 0];% special case, no initial velocities
%
tspan = linspace(0,50,100);
[t2,y2] = ode45(@pend2,tspan,[vo(1) vo(2) tho(1) tho(2)]);
```

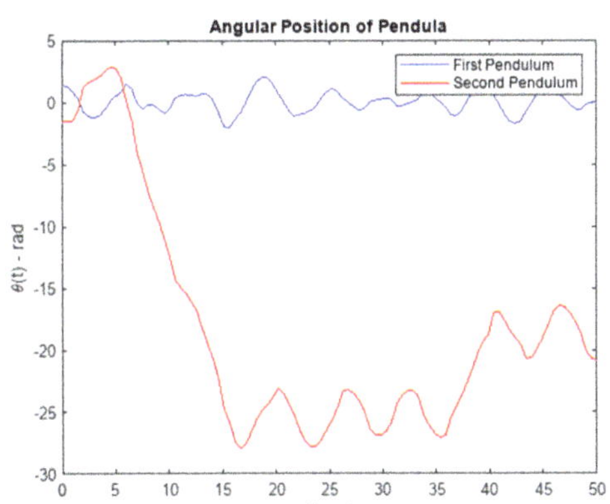

Figure 2.3: Angular location of 2 coupled pendula near a nonlinear displacement.

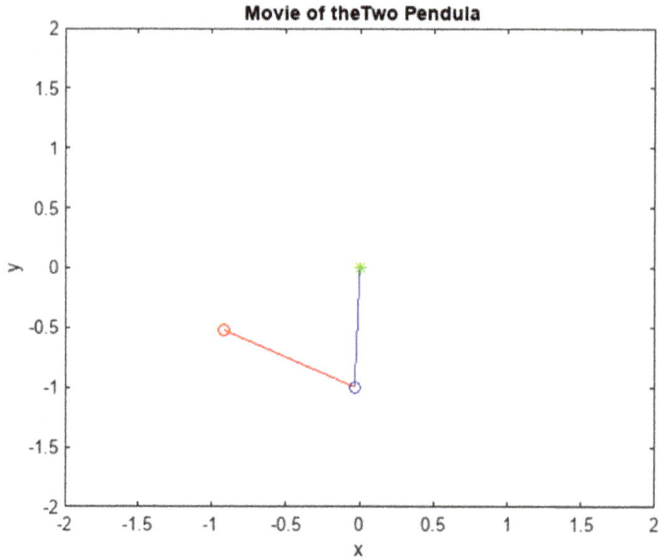

Figure 2.4: Frame of a movie showing the motion of 2 coupled pendula.

2.3 Pendulum and Nonlinear Motion

Normally, small angle motion is assumed for an oscillator such as a simple pendulum. However, for large excursions in amplitude or large velocities, nonlinear behavior is observed. These diverging behaviors typically set in very rapidly, with very small changes in the parameters of the problem. A precursor to instability is the increase in the "phase space" area occupied by the system in $(\theta, d\theta/\mathrm{dt})$ space. For small angles the equation is linear

$$\frac{\mathrm{d}^2}{\mathrm{d}t^2}\theta = -\mathrm{gL}\sin(\theta) \sim -\mathrm{gL}\theta, \quad \omega \sim \sqrt{\mathrm{gL}} \tag{2.2}$$

The initial angle is set to 70 degrees, gL is set to 10, and the user chooses the initial velocity using the "slider" provided. The movie is made for 5 small angle periods or 10 s. As the initial velocity increases the amplitude with respect to the small angle solution increases, as does the period. The increase in the phase space area is also observed in the second plot. The onset of non-linearity is very sharp and occurs at an initial velocity setting of ~297.

```
% Pendulum - Large Oscillations;
% Set initial position and velocity of pendulum
% Initial Angle (degrees)
tho = 70;
tho = (tho .*pi) ./180.0;    % Convert angle to radians
% Initial Angular Velocity (degrees/sec)
vo = 295; % deg/sec
vo = (vo .*pi) ./180.0;
% g/L in MKS units, sqrt(g/L) = w^2, w = 2*pi/T, small angle approx
gL = 10.0;
w = sqrt(gL); %
T = (2.0 .*pi) ./w % small angle period
```

```
T = 1.9869
```

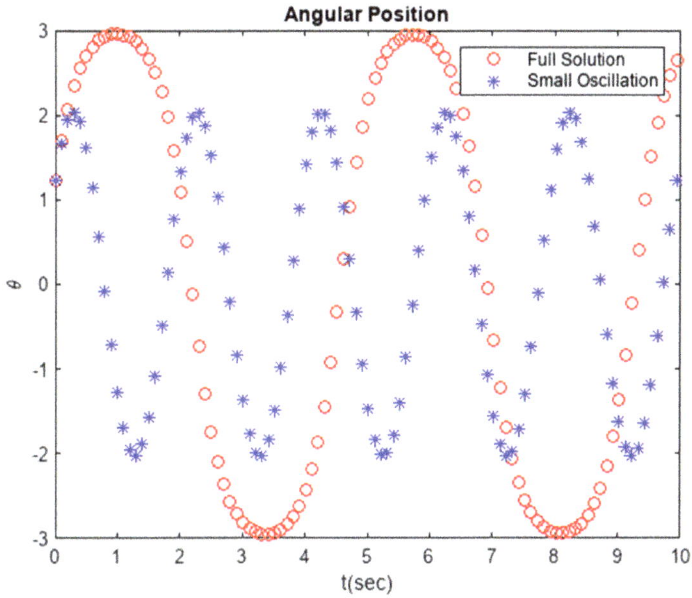

Figure 2.5: Amplitudes for a single pendulum for the linear and non-linear case.

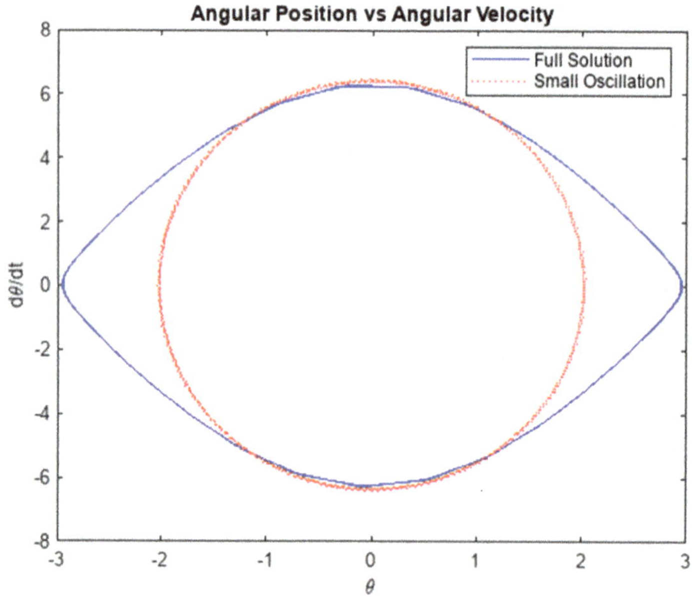

Figure 2.6: The phase space in angle, angular velocity space for a pendulum near non-linear behavior.

2.4 Potential Scattering

One needs to explore how a given force determines an orbit, either bound or unbound. Much can be learned by sending a particle as a probe to scatter off a target. It is the forces acting between the 2 bodies that determine their trajectories. First, the simplest situation is to study the path of a particle in the field of a fixed force center.

2.4.1 *Classical scattering off a fixed force center*

Scattering of one particle off another is one of the methods of exploring the forces between the particles. The scattering off a fixed force center, one which does not recoil, might be an approximation for a very heavy particle that scatters a much lighter particle. The force may be attractive or repulsive — the parameter qq is set by the user. The force is central, directed along the line of centers, with a power law behavior, $F(r) = 1/r^a$ where solutions are found for exponents also set by the user. The solutions are found using the MATLAB package "ode45" for the numerical solution of differential equations. The "impact parameter" b is varied and a "movie" is made of the trajectory. The parameter b is the y value far from the scattering center and with a projectile initial velocity along the x axis at y = b. It is useful to explore the results for different power laws and attractive or repulsive forces. The equations of motion are supplied to "ode45" along with the initial conditions. A movie for different b values is made using "plot".

In the "Command Window" the "Workspace" tab gives all the information on the variables active at the moment. Variable types and values can also by printed to the Command Window with a query at the prompt, ≫, with the variable name.

```
% look at dynamics - trajectory vs impact parameter for
% power law forces 1/r^n, attractive or repulsive, fixed V(r)
% power law exponent a = 1,2,3,4
a = 2;
% qq = 1, -1, attractive or repulsive force, F(r) = qq/r^a
```

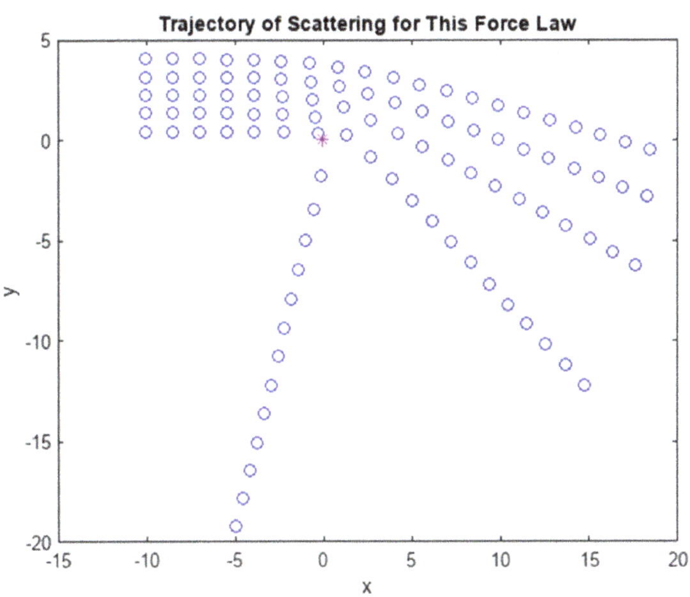

Figure 2.7: Trajectories for scattering with a user chosen power attractive potential.

2.4.2 *Scattering cross section*

In general larger a values mean more localized forces at small r values which leads to larger deflections for a given small value of b, lesser for larger values of b. There are 4 possible choices of a and 2 choices for q, and all 8 plots illustrate the characteristics of this type of scattering. This gives a concrete picture to the idea that the angular distribution is a reflection of the forces between the particles which scatter. Scattering experiments allow for the understanding of the forces involved.

A familiar example of such scattering is Rutherford scattering, where the heavy target nucleus is treated as a fixed scattering center. It is clear that each b has a corresponding angle. We cannot aim at the nucleus, so all transverse area elements are equally probable and the differential cross sectional area is $d\sigma = 2\pi \mathrm{bdb}$. Transforming to the observable scattering angle, $d\sigma = 2\pi b(\theta)(\mathrm{db}/d\theta)d\theta$. Using the solid angle, $d\Omega = 2\pi\sin(\theta)d\theta$ appropriate to azimuthal symmetry the differential scattering cross section is $d\sigma/d\Omega = (b/\sin(\theta))(\mathrm{db}/d\theta)$. The relationship between b and θ in general is $\theta \sim V(b)/T$ where T is the incident particle kinetic energy and V is the interaction potential energy between the projectile and the target. For the Coulomb potential, the Rutherford cross section is approximately derived as shown below. Other power laws rather than Coulomb yield different angular distributions, which indeed is the purpose of the exercise.

$$d\sigma/d\Omega = (b/\sin(\theta))(\mathrm{db}/d\theta) \sim (b/\theta)(\mathrm{db}/d\theta)$$

$$F \sim 1/b^2, \quad t \sim b/v, \quad \mathrm{dp}_y \sim \mathrm{Ft} \sim p\theta, \quad \theta \sim 1/\mathrm{bvp} \sim 2/\mathrm{Tb} \tag{2.3}$$

$$d\sigma/d\Omega \sim \left(1/\theta^2\right)(\mathrm{db}/d\theta) \sim 1/\theta^4$$

Now to 2 body scattering with a target particle which is initially at rest, usually called the "lab frame", but which now recoils when scattered off an incident particle. The target is no longer a fixed point.

2.4.3 *Classical 2 body scattering, target at rest*

The scattering here has no dynamics but is purely the result of the kinematics for the masses $m + M \rightarrow m + M$. All possible scattering angles are calculated, Kinetic energy and vector momentum are conserved in the process. Momentum is assumed to be the non-relativistic value of the incident projectile mass times velocity. The particle labels in the script are 0, T, 1 and 2 for the incident particle, m, the target at rest, M, the outgoing particle and the recoiling target. The projectile mass is, $m = 1$. The target mass is user chosen, as M. The classical equations can be solved by specifying a single variable, for example the target recoil angle ϕ. The target recoil velocity v as a function of the recoil angle of the target is:

$$v = 2\cos\phi/(1 + m/M) \tag{2.4}$$

The final projectile velocity, u, and angle, θ, can also be expressed in terms of the single chosen variable The expression for θ follows simply from transverse momentum conservation.

$$u = \sqrt{1 + v^2} - 2\mathrm{v}\cos\phi, \quad u\sin(\theta) = \mathrm{v}\sin(\phi) \tag{2.5}$$

Note that if $M = 1$ is chosen all the momentum of the projectile can be transferred to the target in a head on collision, as billiards players know well. It is also observed in this case that the angle between the projectile and target is always 90 degrees. In contrast for a large target mass, little velocity is transferred to the target in a collision. This fact in familiar from traffic accidents, say between a truck and a VW. A movie showing the angles and momenta is generated for all possible scattering angles. A heavy target can scatter the projectile backwards. For a light target, the angles are forward. The target mass can be chosen to be less than or more than the fixed projectile mass. The resulting kinematic differences in behavior can be readily and profitably explored.

```
% Display 2 body NR collsions. Target at rest. Kinematics only
% Initialize  - Setup Momentum and Energy conservation
% 0 + T -> 1 + 2 but non-relativistic so mo = 1 = m1, elastic only
% assume projectile velocity is in +x, T is at rest
% target mass M, projectile mass = 1,
M = 1;
% loop on scattering angle in CM
% find recoil velocity and scattered projectile angle and velocity
vo = 3; % initial velocity
% m1/m2 =1/M
```

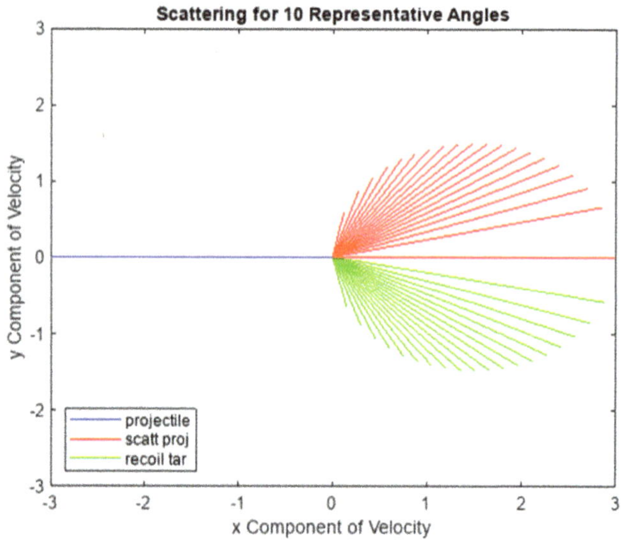

Figure 2.8: Velocity and angle of scattering for a user defined set of masses and incident velocity.

2.4.4 *Relativity, photons and Compton scattering*

There are some changes when relativistic expressions for momentum and total energy are used. An example is Compton scattering, the 2 body elastic scattering of a photon off an electron target at rest. E_o is the incoming photon energy, E the recoil energy and θ the photon recoil angle. The electron, initially at rest, recoils with energy k at a recoil angle ϕ. As with classical elastic scattering with masses fixed, the scattering is defined by the masses, the initial projectile energy and a single recoil angle. The photon is treated as a particle carrying incident energy E_o and momentum E_o/c and scattered energy E.

$$\begin{gathered} E = E_o/[1 + y(1 - \cos(\theta))], \\ y = E_o/(m_e c^2) \end{gathered} \tag{2.6}$$

The photon can always backscatter, for any photon energy. If the photon goes backward, the electron goes forward. If the photon goes forward, the electron goes at 90 degrees. In any case the photon loses energy which means it has a reduced frequency. That means there is a wavelength shift, the famous Compton effect. At low photon energies the angular distribution reduces to that of Thomson scattering, which is forward/backward symmetric with $\sim$ unchanged photon energy. At high energies there is a pronounced forward peak and a large wavelength shift. The Compton wavelength of the electron, mass m_e, is defined to be λ_e and h is the Planck constant. The frequency shift is $\lambda - \lambda_o$.

$$\lambda_e = (h/(m_e c)), \quad \lambda - \lambda_o = \lambda_e(1 - \cos(\theta)) \tag{2.7}$$

The changes in the angular distributions with a change of incident photon energy can be dramatic and can easily be explored using the "edit field" for ω_o. The complete kinematics is deferred until the chapter on "Special and General Relativity". However, if the photon is considered to be a massless particle, it is not surprising that it can always be backscattered by a massive target.

```
% Compton angular distribution as polar plot
% evaluated in target rest frame
% roughly isotropic in cm (1+ct^2) - low energy photons, Thomson
scattering
% wo = is the incident photon energy in MeV
% electron mass in MeV
me = 0.511 ;
% kinematics is 2-body elastic,relation of incident wo to scattered
photon energy w
% NR sigma is Thompson result, 8*pi/3 *(alpha/m)^2
```

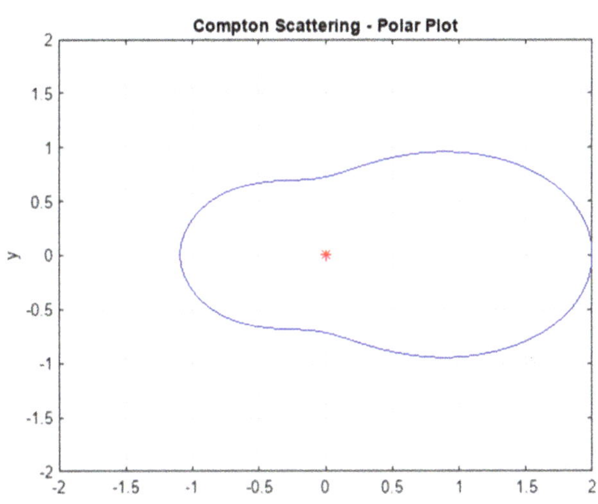

Figure 2.9: Polar plot for Compton scattering at a user chosen incident photon energy.

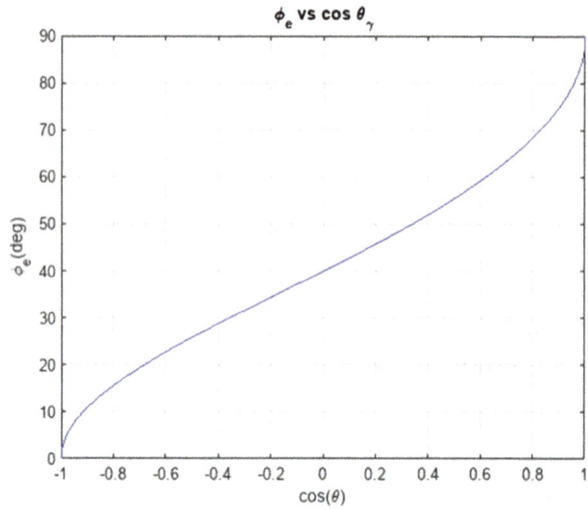

Figure 2.10: Plot of the locus of angles for the scattered photon and recoil electron in Compton scattering.

Backscattered photons have forward going electrons with a finite energy, which depends on the initial incident photon energy. This makes a forward electron energy measurement a useful determination of the initial photon energy.

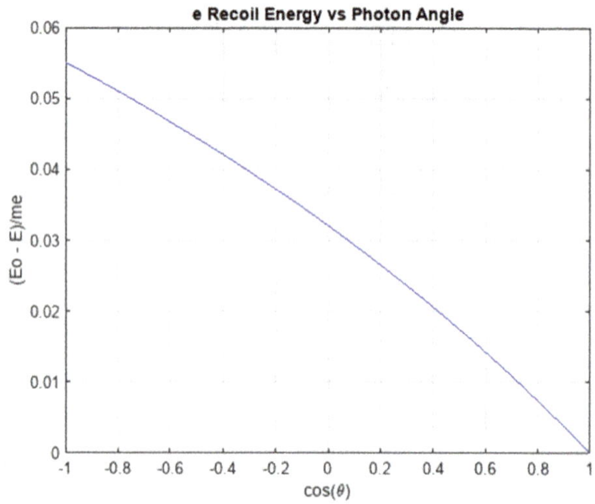

Figure 2.11: Plot of the locus of angle of the scattered photon and recoil electron energy in Compton scattering.

2.5 The Free Subway

A central force, directed along the line of centers, which has an inverse square dependence of the force has interesting properties. First, it is only the matter contained within the radius from a body to the center of a mass like the Earth that has a net force. If the body is inside the Earth's radius, R, the acceleration felt by the body is $a = (r/R)g$ which has the right limits at $r = 0$ and R.

It is an amusing thought experiment to imagine a subway shaft cut along a chord connecting two points on the Earth's surface. Imagine a "free fall": trip along that chord. Because of Gauss's law that the important thing is the vector between the center of a body and the test mass and the matter contained within that radius, there is simple harmonic motion of that test body moving along a chord. For a uniform density sphere of radius R, as an approximation to the Earth, the distance along the chord is x_o, the maximum depth of the subway is d, the angle subtended by the chord is θ, and the distance along the Earth's surface is s, the arc length:

$$\begin{aligned} x_o &= 2\mathrm{R}\sin\theta \\ d &= R(1-\cos\theta), \quad s = 2R\theta \end{aligned} \tag{2.8}$$

The circular frequency of the motion, $\omega_o^2 = G\rho(4\pi/3) = g/R = a/r$, is independent of the particular chord and depends only on constants which define the gravity field of the Earth in the uniform density approximation. The trip time is then half the period, $T = \pi/\omega_o$, which for the Earth is 2533 sec, or about 42 minutes, independent of the trip distance. The code provides a graph of the position vs. time in order to provide insight into the velocity (the slope) as a function of time. The maximum acceleration is reached at the lowest depth, at the half way point in the trip. The velocity is zero at the turning points at the start and end of the trip. Who says there is no free lunch? We can use the gravity field of the Earth to travel for "free." Elon Musk take note! However, the velocity of the trip increases with the distance and the heat encountered in a "deep" trip might be a bit impractical. Still, the study of the use of the properties of the inverse square law for gravity is amusing. The user picks the distance for the trip.

```
% Earth subway - compute free fall through chord of Earth
% Initialize variables, subway is defined by chord
g = 9.81;    % Gravitational acceleration (m/s^2)
re = 6.38 .*10 .^6;  % Earth radius (m)
% pick distance and therefore the chord for the  "free" subway trip
```

```
dist =   843 ; % Subway Distance in km
dist = dist .*1000;
theta = asin(dist ./(2.0 .*re)); % chord
depth = re .*(1-cos(theta)) ./1000 % Free Subway Max Depth (km)
```

```
depth = 13.9386
```

```
sdist = (re .*2.0 .*theta) ./1000 %'Distance Along Earth, arc
length (km)
```

```
sdist = 843.6144
```

```
% Gauss law - uniform Earth density ==> force due to distance to
earth
```

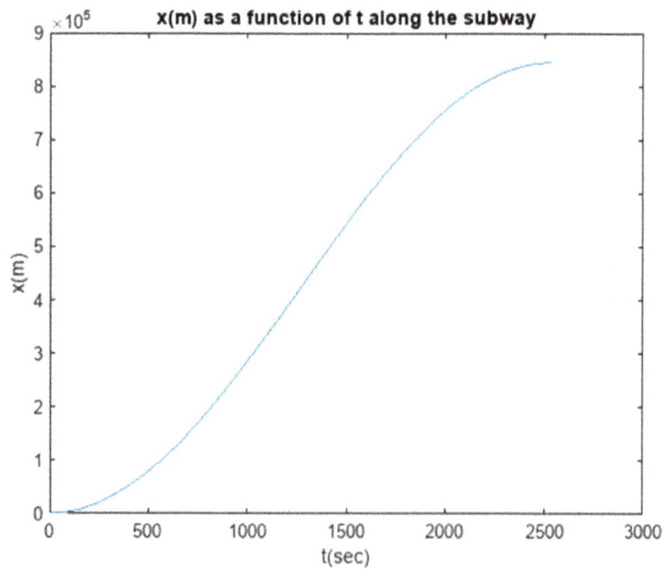

Figure 2.12: Time vs. distance in a "free" subway. The endpoints have 0 velocity.

2.6 Kepler Orbits in Time, Numerical

Before starting the numerical work it is useful to "read" the equation for trajectories near the Sun. Because the force is directed along the line of centers there is a conserved angular momentum, L, and the orbit is in a plane, specified by polar coordinates, r and θ. In particular, a bounded circular or elliptical orbit is "re-entrant" — it repeats in time so that the orientation of the trajectory in the plane does not change from orbit to orbit. This is a unique property of the inverse square law and is not obeyed for other forces. The changes due to General Relativity (GR), the perihelion advance, will be deferred until later.

The first historical answer for the orbits of the planets was that they were perfect circles, as described by Ptolemy. That satisfied the ideal of geometric perfection of the cosmos. Unfortunately it was wrong and Kepler, by observation, not intuition, discovered that the orbits were ellipses. It was the triumph of Newton to show that the inverse square law actually required elliptical orbits in general. Of course, that required that he invent differential calculus, but no big deal for Newton.

Looking at the Newtonian inverse square law, one can conclude that the characteristic time, T, for an orbit should scale independently of the mass of the body and as the 3/2 power of the radius X of the orbit, one of Kepler's laws, in fact. Equating the dimensions of the force and the mass times acceleration, GMm/$X^2 \sim$ mX/T^2, $T^2 \sim X^3$/GM. Numerically, using X of 1 AU, the mass of the Sun and G, the scale of times of motion at ~1 AU from the Sun is ~0.16 years — which is approximately correct. In general, it is a good exercise to read the dimensions of any equation and evaluate some of the parameters in order to be familiar with the characteristic scales associated with the particular force.

The detailed problem is the numerical evaluation of orbits in a central inverse square law. The core computation is done with the use of "ode45", the MATLAB numerical integrator for a set of ordinary differential equations. The initial dialogue sets the distance to the Sun, fixed at 1 AU, and the initial velocity, both radial and angular. In the case considered here the initial radial velocity is zero in the interest of simplicity, which is a turning point of the orbit.

The motion defined by these initial conditions is displayed as a (x, y) movie. The movie time spacing is uniform, so that an intuition for orbital velocity can be attained by trying different configurations. Radial velocity on a given orbit is largest nearest the Sun, as expected. Results for an initial radial velocity = 0 and tangential velocity in AU/yr of 6.28, 8 and 9.5 in the specific case of an initial radius of one astronomical unit, or AU, can be executed. An AU is the mean Sun to Earth distance. The velocities then correspond to the possible Kepler orbits of a circle, an ellipse and a hyperbola. For the circular orbit, the Sun is at the center, while for the elliptical, the Sun is at one of the foci of the orbital ellipse. Other shaped orbits are possible and the user should explore the possibilities. For example, a parabola occurs at the transition point from bound to unbound orbits. If the velocity is less than that for a circular orbit, the minimum for a stable bound orbit, the trajectory starts to spiral in an decreasing radius. The time scale for integration is the time span of five circular orbits at the starting radius. Note that the numerical results do not appear to be exactly re-entrant, due to the ode45 code, but this is simply numerical "noise" and should be recognized as such.

```
G = 6.67e-11; % MKS units
Mo = 2.0e30;    % solar mass
au = 1.49e11;  % AU = earth-Sun distance, m
yr = 60.0 .*60.0 .*24.0 .*365.0;    % sec
vc = sqrt(G .*Mo ./roa); % circular velocity
T = (2.0 .*pi .*roa) ./vc ; % circular period
```

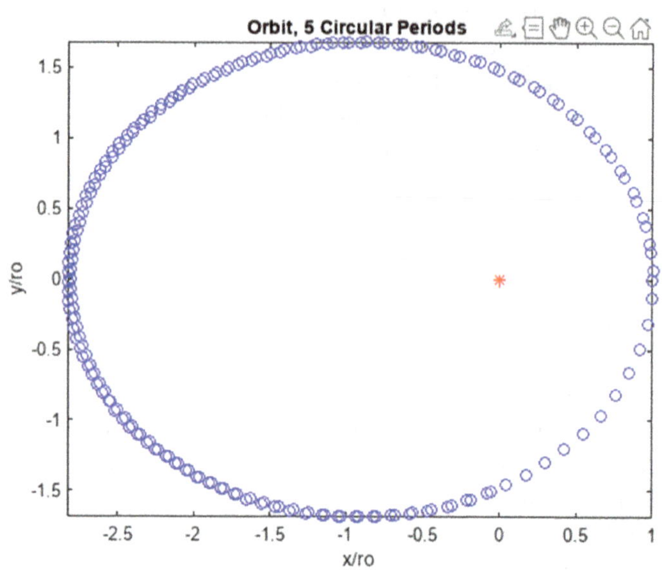

Figure 2.13: Movie frames of a closed orbit. Velocity is greatest nearest the Sun.

2.7 Kepler Orbits and Energy

Orbits labelled by time, found numerically, show when the motion is slow, far from the Sun, or when it is rapid, nearer the Sun. It is also clear that the orbits repeat in a Newtonian inverse square law. However, the energy formulation of orbits gives additional symbolic insights, although the time information is largely lost. Since gravity is a central force, the motion is in a plane and there is a conserved angular momentum, $L = \mathrm{mr}^2(d\theta/\mathrm{dt})$ where θ is the polar angle in that plane and r is the radius. The velocity of a circular orbit, v_o, at a radius, r_o, equating the centrifugal force and the gravitational attraction is $v_o = \sqrt{\mathrm{GM}/r_o}$. There is no dissipation for the planets, so a conserved energy E can be defined with both kinetic and potential terms. That constraint leads to an effective 1 dimensional motion.

$$E = m\left(v_r^2 + v_\theta^2\right)/2 - \mathrm{GMm}/r = T + V, \quad v_\theta = r(d\theta/\mathrm{dt}),$$
$$L = \vec{r} \times \vec{p} = \mathrm{mr}^2(d\theta/\mathrm{dt}) \tag{2.9}$$

The effective 1 dimensional potential is scaled with respect to a circular orbit of radius r_o. The circular orbit has the minimum energy of all the orbits. That radius has a velocity, $v_o = \sqrt{\mathrm{GM}/r_o}$ with an angular momentum, purely tangential, of $L_o = m\sqrt{\mathrm{GM}r_o}$. The effective energy, scaled to the orbiting mass m, is then simply;

$$U = E/m = (\mathrm{dr}/\mathrm{dt})^2/2 + L^2/(2m^2r^2) - \mathrm{GM}/r = \mathrm{GM}(-1/r + r_o/2r^2),$$
$$U_o = \mathrm{GM}/2r_o \tag{2.10}$$

The added term in U is the centrifugal potential, which scales as the inverse square of the radius. The dimensionless energy variable for an orbit of energy U is the ratio of U relative to a circular orbit with $q = -1$. A circle has $q = -1$, a parabola $q = 0$ and an hyperbola $q > 0$ (unbound motion). Positive energy indicates unbound trajectories, while negative energies are bound.

$$q = U/|U_o| \tag{2.11}$$

The orbital radius is $r/r_o = 1/(1+\mathrm{e}\cos(\theta))$. The eccentricity, e, major axis, a, and the turning points for a bound orbit are all expressed in terms of q:

$$e = \sqrt{1+q}, \quad a = 1/|q|, \quad x_{\mathrm{turn}} = (-1 \pm e)/q \tag{2.12}$$

The velocity with respect to a circular orbit with e $= 0$ is $v_r/v_o = \sqrt{q + 2(1 + \text{ecos}(\theta))}$. The code below has an initial radius of 1 mean distance from the Earth to the Sun, 1 AU. The user chooses the value of q and the resulting energy and orbit is plotted. The turning points, x_1, x_2 are indicated on the potential plot, and the near elliptical focus, a, is shown along with the ellipse and the major and minor axes of the ellipse. The period of the orbit in years is printed. The user chooses q and all types of orbits can be generated.

```
G = 6.67e-11; % MKS units
Mo = 2.0e30;    % solar mass, kg
au = 1.49e11;   % AU = earth-Sun distance, m
yr = 60.0 .*60.0 .*24.0 .*365.0;    % year in sec
ro = 1.0; % AU
% Find Effective 1-d Potential - Use centrifugal Potential, V~1/r^2
xxx = linspace(0.25,10.0);    % r variation in ro units
Ueff = 1 ./(2.0 .*xxx .*xxx) - 1.0 ./xxx;
Ueff = Ueff .*2;
```

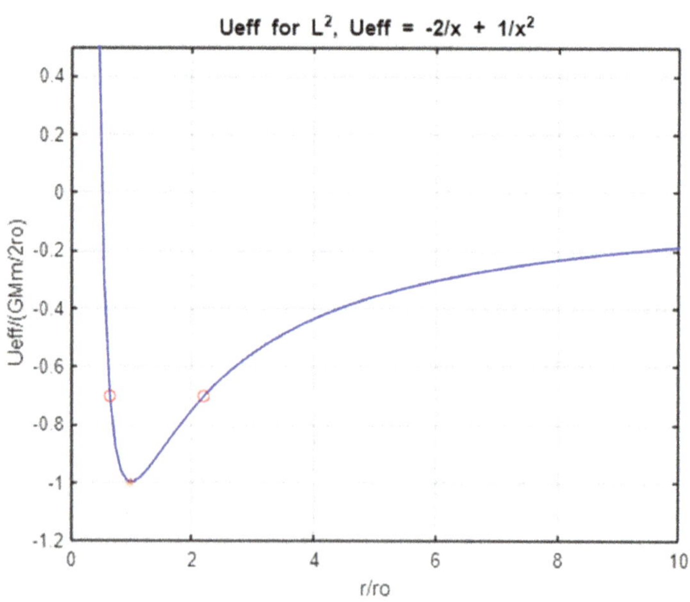

Figure 2.14: Effective potential. The minimum (red star) is for a circular orbit.

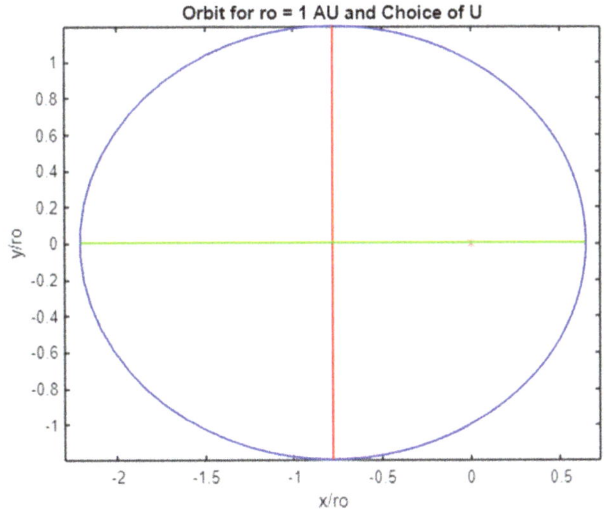

Figure 2.15: Elliptical orbit for orbit shown as red circles in the previous Figure. The Sun is at the semi-major axis (red star).

2.8 Basic Rocket

Turning from Kepler orbits to the rockets that enable exploration of the planets moving in those orbits, consider rocket motion as a prelude to the exploration of other planets, Start with the simplest rocket, no gravity well to escape, and no air drag. Use momentum conservation in 1 dimension, assuming a purely vertical launch. The exhaust velocity of the fuel with respect to the rocket is v_o where m is the present rocket mass, which gets a velocity increase dv by ejecting a mass dm with velocity vo.

The velocity as a function of rocket mass, initial value m_o, is found by integration taking the initial velocity to be 0.

$$\begin{aligned} \text{mdv} &= -v_o \text{dm} \\ v(m) &= v_o \ln(m/m_o) \end{aligned} \tag{2.13}$$

Assuming a constant rate of fuel consumption, dm/dt, with no payload, $m(T) = 0$, and the total burn time is $T = m_o/(\text{dm/dt})$. Solve symbolically for v(t) and then integrate to get y(t). A final payload mass can be taken into account using a reduced burn time. The payload mass is a user chosen parameter.

It is notable that the final velocity depends only logarithmically on the mass ratio, but directly on the exhaust velocity. That puts a premium on fuel with the highest exhaust velocity. Plots are made of the velocity and distance as a function of time. The payload fraction of the total mass can be varied and a larger payload means a shorter burn time and a lesser final velocity, as seems reasonable.

```
% solve the rocket equation - free of forces, drag
% d2y/dt2 = vo /(T-t),vertical y, exhaust velocity vo,T = total
burn time
%
syms t T Dy y(t)
Dy = diff(y,t);
odev = Dy-vo/(T-t) == 0;
vs = dsolve(odev,'y(0)=0')
```

vs = $vo \log(T) - vo \log(T - t)$

```
ys = int(vs,t);
y = simplify(ys)
```

y = $-vo\,(T - t)(\log(T) - \log(T - t) + 1)$

```
% total possible burn time T is mo/(dm/dt)
% payload ratio  mp/mo = 1-tp/T, tp = burn time for this payload
% Payload ratio mp/mo => Payload burn time tp = T(1-mp/mo)
% vo = exhaust velocity, acceleration in vo/T units, velocity in vo
units  \n')
```

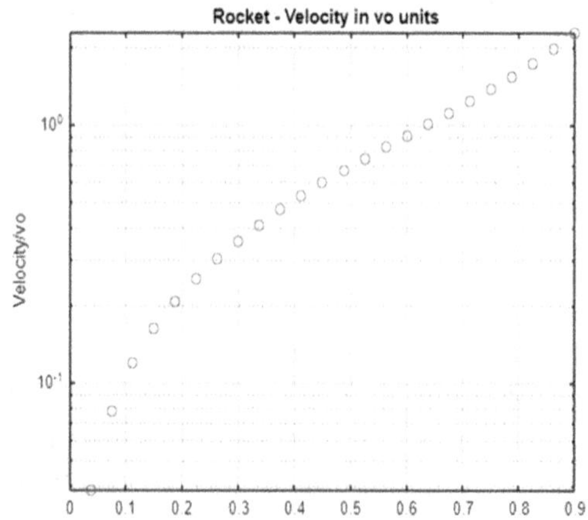

Figure 2.16: Movie of a rocket trajectory as a fratcion of the total burn time. The maximum acceleration is at the end of the "burn".

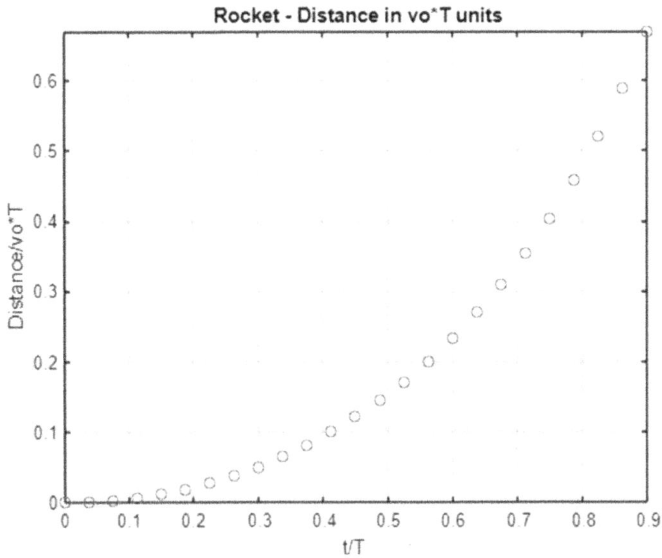

Figure 2.17: Distance travelled as a function of the fraction of the total burn time, T.

2.9 Rocket on Earth (g) with Air Drag

A rocket is launched from Earth. Look first with a symbolic solution at an approximate payload percentage for Earth orbit and Earth or Sun escape velocity. Then use MATLAB ode45 numerical integration to take into account a launch with g from Earth's surface, g falloff with altitude, as well as the viscosity (drag) of the atmosphere, whose density falls off approximately exponentially with altitude. Note that for the numerical calculation, there is zero payload assumed. The rotational velocity of the Earth is ignored. However, real launches use that velocity and most launch sites are located near the equator in order to utilize the added rotational velocity.

Escape velocity is v_e calculated by equating kinetic energy with depth of the gravitational well, $mv_e^2/2 = \mathrm{GMm}/r_e$. Orbital velocity, calculated equating the $\sim$ mg force with the centrifugal force in circular motion, $v_{\mathrm{orb}}^2/r_e = g$ is less than the escape velocity by a factor $\sqrt{2}$, assuming low Earth orbit. The payload ratio reduces the final velocity exponentially reflecting the logarithmic dependence of the velocity on mass. Note the extremely small payloads if one wishes to escape from the Sun. Another strategy is needed. The deceleration due to viscosity acts proportionally to the square of the velocity.

The falloff of the Earth's deceleration is taken into account with the modification, $g \to g(R_e/(R_e + y))^2$ where R_e is the Earth's radius.

$$m(t) = m_o - (\mathrm{dm/dt})t, \quad \rho = \rho_o e^{-\beta y},$$
$$\frac{\mathrm{d}^2}{\mathrm{d}t^2} y = v_o(\mathrm{dy/dt})/m(t) - g(R_e/(R_e + y))^2 - \kappa\rho v^2/m(t) \tag{2.14}$$

The user selects the exhaust velocity and the payload mass ratio. These settings can be varied to attain low Earth orbit or escape velocity from the Earth. Three scenarios are explored; a free rocket, a rocket in the Earth's field, and a rocket in the Earth's field impeded by the viscosity of the atmosphere. All 3 trajectories are plotted showing both the altitude and the velocity as a function of time.

```
% Solve non-relativistic rocket, numerically using the old Saturn V
as an example
% veq = (2.0 .*pi .*re) ./(24 .*3.6e3);    % equatorial launch
velocity km/sec
% not used here, but launch locations are near the Equator (Florida)
 to use the rotational velocity boost
rs = 1.5e11;  % distance to sun - m
me = 6.0e24;  % earth mass - kg
ms = 2.0e30;  % sun mass, - kg
vorb = sqrt(g .*re)   % orbital velocity - circular, low orbit
~ 7.9 km/sec
```

```
vorb = 7.9060e+03
```

```
ve =sqrt(2.0 .*g .*re)  % escape velocity for Earth ~ 11.2 km/sec
```

```
ve = 1.1181e+04
```

```
vs = ve .*sqrt(ms .*re ./(me .*rs)) % escape velocity to leave
solar system ~ 42 km/sec
```

```
vs = 4.2093e+04
```

```
% air density falloff with altitude - m
zo = 9140;
dragg = 0; % for now a simplification
(exp(-vs ./vo)) .*100  % estimated payload  in % for free rocket to
attain solar escape velocity
```

```
ans = 0.0106
```

```
(exp(-ve ./vo)) .*100  % estimated payload  in % for free rocket to
attain earth escape velocity
```

ans = 8.7985

```
(exp(-vorb ./vo)) .*100 % payload in % to attain orbital velocity
```

ans = 17.9301

```
% accel = vo*(dm/dt)/m-g*re/(re+y)-visc*v^2*e(-y/yo)/m
% vo is the simple rocket equation,g is the Earth drag with falloff
with
% altitude.  visc is the atmospheric viscosity scaling as v^2, but
falling
% off with altitude.
```

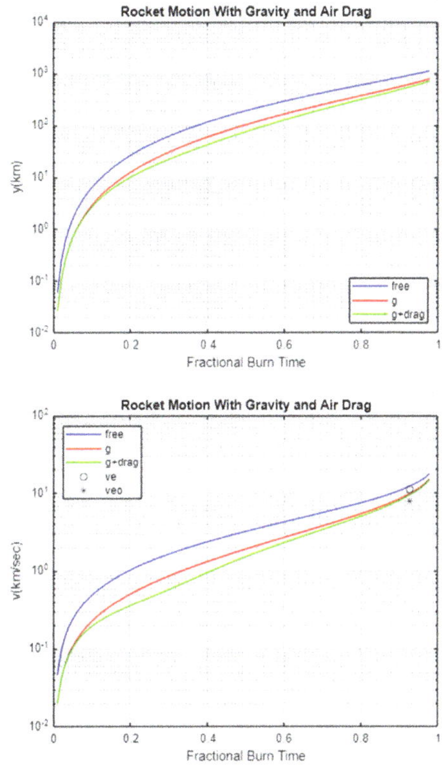

Figure 2.18: Top — distance as a function of the fractional burn time. Bottom — velocity vs t/T. Three cases are plotted; a free rocket, a rocket in a gravity field g, and a rocket with the addition of air drag.

2.10 Reentry to Earth

There is the question of how to return to Earth once in low Earth orbit or further away. It is assumed that the descent is unpowered and that the atmosphere will be used to brake the descent. One issue is that the air is compressed because it cannot get out of the way of the descent fast enough since the descent speed from low Earth orbit, about 8 km/sec, is initially much greater than the speed of sound which is 0.32 km/sec at sea level and standard temperature and pressure, STP. That means the air will heat up, much like the situation when a bicycle tire heats up when being pumped to higher pressures.

This is a complex process, which can be approximated, for a first look, as a process of adiabatic compression. The heat flux, in W/m^2, is F_H and depends on the velocity of the descent object, v and the density of the atmosphere, ρ, as $F_H = \rho v^3$. This type of strong dependence on velocity was already seen in the drag due to atmospheric viscosity during ascent. As a numerical example, take a 1000 kg (1 Tonne) descent module moving at LEO velocity. The kinetic energy of the module is initially 6.4×10^{11} J or 640 GJ. The atmospheric falloff with altitude is assumed, as previously, to be exponential with scale 9.14 km, so that the density at 100 km LEO is $2.1 \times 10^{-5}\,\mathrm{kg/m^3}$. The heat flux is then 5.4 $\mathrm{MW/m^2}$ which needs to be dissipated using a "heat shield" which basically is ablated and protects the module itself. For comparison one ton of dynamite releases only 4.2 GJ of energy so this is a delicate maneuver.

The kinematics of the descent are explored in the code below. The thermodynamics are complex and will not be further explored. Only the trajectory is examined. The order of magnitude of the heat flux is daunting enough. The rocket ascent was powered and was essentially a one dimensional problem, assumed to be fully vertical. The descent is unpowered and, by necessity, is a two dimensional problem. The acceleration during descent has a term due to the Earth's attraction and a drag term similar to what was seen during ascent.

$$\frac{\mathrm{d}^2}{\mathrm{d}t^2}\vec{r} = -(\mathrm{GM})(\vec{r}/r^3) - [\rho(y)C_D(y)Av\vec{v}]/2 \qquad (2.15)$$

The drag term represents the atmosphere with a density, $\rho(y)$, which falls off exponentially with altitude and a term which depends on the Mach number, M, which is defined to be the ratio of the speed of the object to the speed of sound. This behavior is simplified in the code, but the coefficient still peaks at M of one. The drag deceleration depends on the area, A, of the object, points in the direction of the velocity and is proportional to the square of the velocity.

The two dimensional equations are solved numerically using the Matlab utility "ode45". The value for the LEO altitude is fixed at 100km and the angle of descent at 20 degrees. Since the thermal issues are being ignored, the angle is not very sensitive, but in reality it is a parameter that needs to be finely tuned as several descent tragedies have illustrated. The results of a particular choice of parameters are displayed below. The curves of altitude vs. velocity and altitude vs. deceleration are very characteristic of an unpowered descent.

The descent starts with a rapid loss of altitude but little loss of velocity since the atmosphere is attenuated at 100 km. At an altitude of about 50 km and below the loss of altitude with velocity is roughly linear. The final conditions, $v = 0$ and $y = 0$ are not imposed, since the behavior is entirely determined by the initial conditions in this simplified model. In practice small rockets exist to make adjustments so that it is not simply an initial value problem but a problem in piloting. The deceleration is small initially and then peaks at an altitude of about twenty km to a value of almost 20 "gees". The peak deceleration value depends on the initial altitude. After the peak, the falling value of the velocity shows that the deceleration falls rapidly to zero at lower altitudes. Typically the landing is enabled using parachutes to come to rest on either land or a body of water. These techniques give the pilot a margin for error

```
% look at re-entry from low earth orbit. Vary height and angle of
descent
% solve for the motion - in a plane, y = height, x perpendicular
% -G*Me*y/r^3-(rho(y)C(Ma)*A*v*vy)/2 ; acceleration
% find Mach numbers, velocity w.r.t. sound speed
```

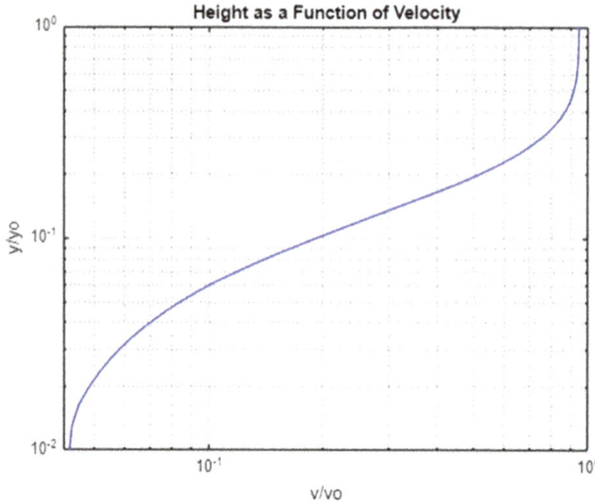

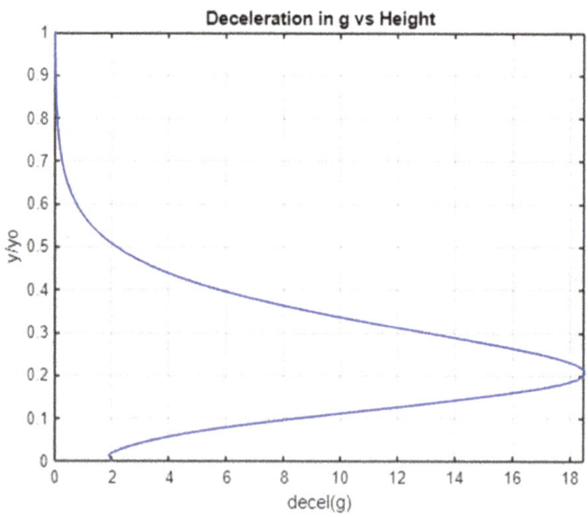

Figure 2.19: Capsule reentry into Earth's atmosphere. Top — velocity as a function of altitude. Bottom — deceleration as a function of altitude.

2.11 Foucault Pendulum

A Foucault pendulum is one installed on the Earth's surface which thus exists in an accelerated coordinate system due to the Earth's rotation. The fact that the Earth is rotating, and thus accelerated, relative to a system of absolute rest, according to Newton, means that there exist "fictitious" forces. One possible way to think about it is that the pendulum oscillates in Newton's "absolute space" while the Earth rotates under it. In the Earth frame the equations of motion have an added term due to the Coriolis force, where Ω is the rotation frequency of the Earth and φ is the latitude. The effect clearly vanishes at the poles. The effect depends on the velocity of the pendulum and repeats on a daily basis.

$$\frac{\mathrm{d}^2}{\mathrm{d}t^2}x = -\omega^2 x + 2\Omega \sin\phi(\mathrm{dy/dt}), \quad \frac{\mathrm{d}^2}{\mathrm{d}t^2}y = -\omega^2 y - 2\Omega \sin\phi(\mathrm{dx/dt}), \tag{2.16}$$

In the absence of rotation the equations of motion describe simple harmonic motion in the (x, y) plane of the Earth's surface. With the rotation, the motion in x and y is coupled due to the terms proportional to velocity. The symbol ω refers to the frequency of the pendulum in the absence of accelerated motion.

These equations are solved symbolically using the "dsolve" MATLAB utility. The approximate solution for the case where ω, the pendulum harmonic frequency, is much larger than Ω is also shown for comparison. The script is quite slow because at least an entire day is mapped out; be patient. In addition, the exact x and y positions have extremely complex solutions which are not printed here. They are displayed in the Earth frame for the case where the pendulum period is 0.1 days and it sits at thirty degrees north latitude. The approximate solution is also plotted by the script as a movie and illustrates that the motion is not fully re-entrant on a daily basis in this case because of the approximations made in the numerical solution found by "ode45".

The approximate solution factorizes into simple harmonic motion times a modulation of the amplitude due to the Coriolis force.

$$\begin{aligned} &\Omega' = \Omega \sin\phi \\ &x = \sin(\Omega' t)\sin(\omega t), \quad y = \cos(\Omega' t)\sin(\omega t). \end{aligned} \tag{2.17}$$

The generated plot shows the exact solution in the pendulum frame in blue and a view in the "absolute" frame in red where the pendulum exhibits simple harmonic motion. A movie of the two motions is plotted. The motion repeats daily.

```
% precession of a Focult pendulum
% k1 is natural frequency = w^2, k2 is 2*omega*sin(phi)
% phi is latitude, omega is Earth frequency, pick time in days,
omega = 2*pi
% pendulum precesses by 2*pi*sin(phi) per day, x is East and y is
North in Earth system
% D2x = -w^2x + (2*omega*sin(phi))*dydt
% D2y = -w^2y - (2*omega*sin(phi))*dxdt
% the exact and explicit solutions are very complex and will not be
displayed here.
% the long time for the movie to begin and run is a reflection of
the complexity of the exact solutions
```

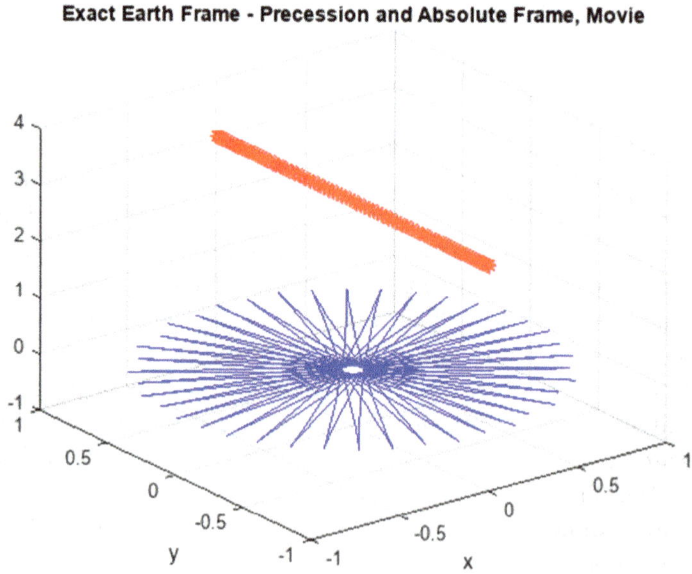

Figure 2.20: Plot of the pendulum position, in the exact case, on Earth (blue) and in absolute space (red stars) as the last frame of a movie of the Foucault motion.

2.12 Euler Angles

The Euler angles were created to relate the position of a rigid body in its own rest frame (body frame) to the position in a frame where the body is in motion (space frame). A special case was quoted in the first section for the situation relating the 2 dimensional (θ, ϕ) center of mass frame to the lab frame. In that case a vector was transformed from a frame to another rotated by the spherical polar coordinates.

The general transformation applies to an extended body. The transformations first rotate about the z axis, followed by a rotation about the x axis and then a second rotation about the z axis. The cosine and sin, c and s, of the three angles are indicated below. In the generated Figure, all angles are 60 degrees to make the visualization easier.

$$\Longrightarrow \begin{bmatrix} c & s & 0 \\ -s & c & 0 \\ 0 & 0 & 1 \end{bmatrix} \Longrightarrow \begin{bmatrix} 1 & 0 & 0 \\ 0 & c_1 & s_1 \\ 0 & -s_1 & c_1 \end{bmatrix} \Longrightarrow \begin{bmatrix} c_2 & s_2 & 0 \\ -s_2 & c_2 & 0 \\ 0 & 0 & 1 \end{bmatrix} \tag{2.18}$$

A visualization of the effect of these rotations is provided by the code which appears below. The locations on a circle and the x and z axes are indicated by blue −> red −> green −> black for the initial axes and then the results of the three successive rotations. The first rotation is in the (x, y) plane, while the second is about the transformed x axis. The final third rotation is about the transformed z axis. The specification of the position of a rigid body requires three angles in Euclidian three dimensional space, assuming the C.M. velocity is zero.

The use of Euler angles will be illustrated by the study of top motion. However, they are perfectly general and can be applied to any situation describing rigid body motion.

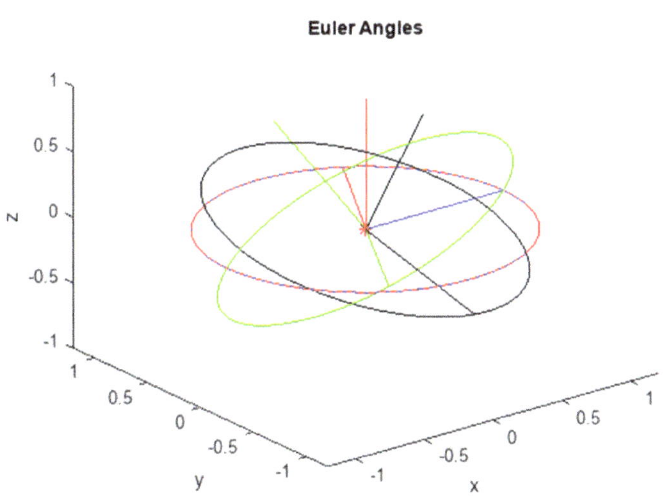

Figure 2.21: Axes for the 3 sequential Euler angle transformations.

Euler angle rotations from body to space frames show the x and z axes and a circle in the (x, y) plane for the three successive rotations, blue is initial, red is after the first rotation, followed by green and then black. The orientation is indicated by the x and z axes and a circle in the (x, y) plane in each case.

2.13 Top — Precession and Nutation

A simple initial example of rigid body motion is that of a top spinning without external forces acting on it. The shape of the top is defined by the moment of inertia about the rotation axis, I_z, and the moments about the other principle axis, I_1, assuming that the top is rotationally symmetric. The top axis of rotation precesses in this case with a frequency ω_p where ω is the top rotational frequency and the top axis is inclined by the angle θ with respect to the vertical, which remains constant. The precession frequency depends on the top being non-spherical.

$$\omega_p = [(I_z - I_1)/I_1]\omega \cos(\theta) \tag{2.19}$$

An illuminating use of the Euler angles is the study of top motion where the top is acted on by a torque due to gravity which then changes the angular momentum of the top. The Euler angles are shown below.

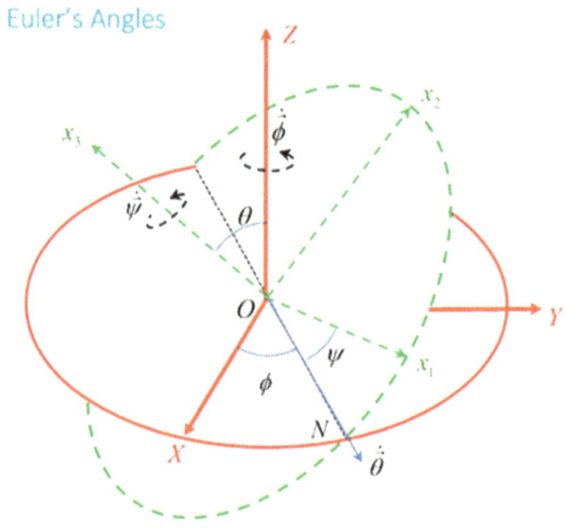

Figure 2.22: Euler angle definitions for the top example.

The top is defined by 3 Euler angles, θ and ϕ define the rotation axis. The angle ψ defines the rotation about the spin axis. There is a second order differential equation for the 3 angles which means that 6 initial conditions are needed. Due to rotational symmetry (like L = angular momentum for a point particle), there is a constant of motion ω_s. There remain five variables to solve for, for the three Euler angles specifying the top position and their derivatives, less the constant of motion, ω_s. The torque is defined by the mass of the top, m, and the height of the C.M. of the top above the fixed point of rotation, h.

$$\omega_s = 3/2 + \sqrt{3I_1 \mathrm{mghcos}\theta}\, I_z \tag{2.20}$$

In general the top rotational axis precesses, as in the forceless case, although the precession rate is now not a constant. It also nutates, with a complex path swept out in the polar angle variable. This motion has the top initially dropping under gravity and then rising in a periodic fashion. In fact, with extreme values of the parameters, the top will even "tumble" and then right itself later. The second order equations to be solved in the space frame are:

$$\begin{aligned} d\psi/\mathrm{dt} &= \omega_s - \cos\theta(d\phi/\mathrm{dt}) \\ \frac{\mathrm{d}^2}{\mathrm{d}t^2}\phi &= [(I_z\omega_s - 2I_1\cos\theta d\phi/\mathrm{dt})d\theta/\mathrm{dt}]/I_1\sin\theta \\ \frac{\mathrm{d}^2}{\mathrm{d}t^2}\theta &= [\mathrm{mgh} - (I_s\omega_s - I_1\cos\theta d\phi/\mathrm{dt})d\theta/\mathrm{dt}]/I_1 \end{aligned} \tag{2.21}$$

The three variables are the polar and azimuthal angles of the top axis in the space frame, ϕ, θ, and the angle ψ defined by the spin of the top itself about its axis of rotation. This set of equations is solved numerically below using the MATLAB utility "ode45". The initial azimuthal angle ϕ is set to zero, while the initial polar angle θ, which defines the inclination of the top, is defined by the user. The initial value of the time rate of change of the azimuthal angle sets the behavior of the nutation, and it is also a user supplied input. The precession rate is not constant, as in the forceless case, but is modulated by the nutational motion. The axis "nutates" and is no longer at a fixed value of θ. The top initial conditions are simplified to initially drop under gravity which is followed by a rise under a restoring torque.

If the time rate change of initial azimuthal angle is set to zero, the top merely drops initially in polar angle until it rises and executes simple harmonic motion. There is another special case, where the top is initially vertical, the "sleeping top", where the top then simply spins about the z axis as in the forceless case because the torque is then zero. The motion is complex and the user is encouraged to vary the top defining parameters and see what the subsequent motion is. The top may "sleep", or simply precess, or fall below $z = 0$.

```
% now setup the top motion
% Top motion depends on spin, Top moments of I, mass \n')
% and initial posiition and velocity (4 of) \n')
% Special Case of "drop", theta = phi = phidot = 0, only thetao \n')
%
```

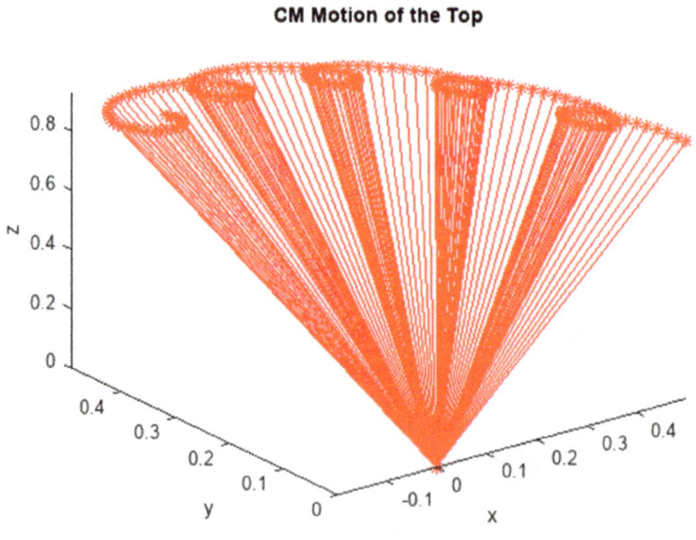

Figure 2.23: Top motion for a user defined set of initial conditions. Both overall precession and nutation are clearly seen.

2.14 Random Walk — 2d

This code uses the MATLAB random number generator utility "rand" to look at a random walk in 2 dimensions begun at $(x, y) = (0, 0)$. The number of steps, N_s is chosen by the user. Each step is of length 1 in a random direction. Typically random deviations have a mean value which scales as the square root of the number of steps. This behavior is observed here, with statistical variations due to the limited number of steps. The actual path is quite illuminating, both the actual steps and the cumulative distance.

```
% Look at a random walk in 2-d
Ns = 100; % number of steps, deviation ~ sqrt(Ns)
```

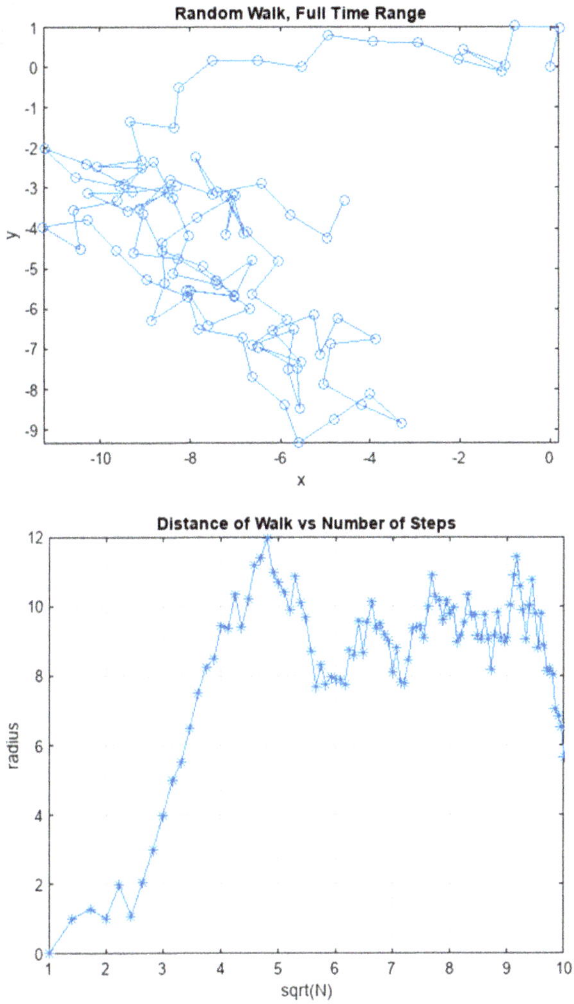

Figure 2.24: Results of a numeric Monte Carlo random walk in 2 dimensions. Top — the actual path in x and y. Bottom, the cumulative distance from the origin showing square root behavior.

2.15 Malthus

The evolution of a population as a function of time is an interesting problem to make a model of — for predictive purposes. The topic is not strictly "Mechanics" but is a relevant one. Here a very simplistic model is explored. Malthus noted that If the population increases proportional to the population, the population increases exponentially with time, soon exhausting the food supply. This observation used to give people pause. However, one way to evade this conclusion is to note that the food supply can increase due to improved farming methods (fertilizers, intensive farming, etc.). Another possible mitigation would be that the population increase could fall with fewer births per parent, a situation now obtaining in several countries.

A simple minded model is explored here which uses 2 equations to make predictions. A food supply c, grows linearly with time governed by a parameter b, but it also falls due to the nutritional needs of the population, −aN. The population then has both growth and reduction factors.

$$\mathrm{dN/dt} = \mathrm{aN}(t), \quad N(t) = N(0)e^{\mathrm{at}}$$
$$\text{or} \tag{2.22}$$
$$\mathrm{dc/dt} = -\mathrm{aN} + b, \quad dN/\mathrm{dt} = -b(N - c)N/c$$

Unlimited growth is impossible for any model of food supplies which does not assume their unsustainable exponential growth. Population will always fall in the long run, else the Earth's surface would consist entirely of bacteria. This exercise gives "food for thought". The code can be changed by the user to vary both the a and b parameters, with the initial values of N and c fixed. There is no closed form symbolic solution for these equations so a numerical solution is adopted here.

```
% MALTHUS - poplation model
% dN/dt = -b(N-c)N/c, dc/dt = -aN+b, c = "food", -aN eaten up,
b = better
% farming, dN/dt -> bN/c, growth exp with <t> = c/b, or not <t> =
c/(bN)
```

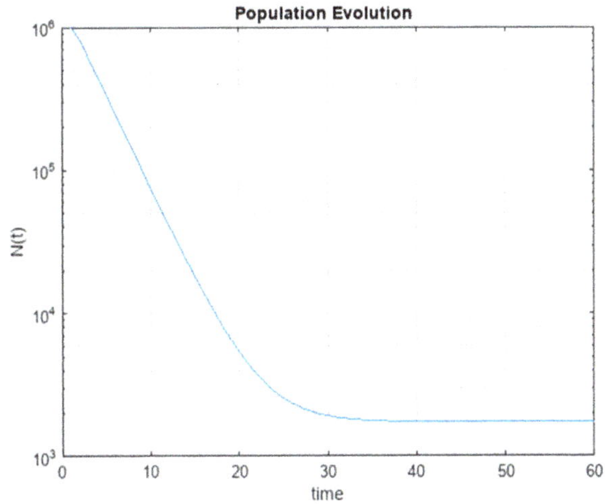

Figure 2.25: Population as a function of time in a simplified Malthusian model. Note the steady state drop by a factor of ~500.

Chapter 3

Electromagnetism

"One day sir, you may tax it." Faraday's reply to William Gladstone, then British Chancellor of the Exchequer, when asked of the practical value of electricity.

"Little you know the subtle electric fire that for your sake is playing within me." Walt Whitman

"Magnetism, as you recall from physics class, is a powerful force that causes certain items to be attracted to refrigerators." Dave Barry

The problems looked at in this section are of 2 main types. The first is the dynamics of motion in an electromagnetic field. The second concerns the static solutions for fields caused by charges and currents. After gravitation, where the constants of coupling and masses, GMm, acting as an inverse distance squared for between masses M and m with a "coupling constant "G", electromagnetism is similar. The charges are discrete and are all multiples of the fundamental charge e, so there is a factor proportional to the square of e in the force. The coupling is then $q_1 \mathrm{e} \mathrm{q}_2 e/(4\pi\varepsilon_o)$, where ε_o is the electrical permittivity of the vacuum, $= 8.85 \times 10^{-12}\ F$. Comparing the electromagnetic force between 2 electrons and the gravitational force, the ratio is 4.15×10^{42}. These scales explain why electric forces are the most commonly used at present, being so much stronger than the forces of gravity.

One can attempt to "read" the equation as was done for the Schrodinger and gravitational equations. For solar orbits G was quite weak and the distances were large, $\sim$ AU. Atomic scales are much smaller. Now consider a simple atom with an electron orbiting a proton. For a circular orbit, knowing the characteristic size is ~0.1 nm, the velocity is 0.005c which is comparable to the electromagnetic fine structure constant, $\alpha = e^2/\hbar c$, and the orbital time is 3.9×10^{-16} s. With the stronger electromagnetic forces the speeds are larger and the times are reduced. Later, in the section on quantum mechanics these order of magnitude estimates for velocity will be found to be in reasonable agreement with the Bohr model of hydrogen.

The scripts in Section 1 were shown fully. For this and subsequent Sections, the scripts will be heavily reduced in length, but are available in the MATLAB scripts themselves. This edited reduction is made in the interest of the length of the volume. In addition many of the techniques to solve a problem and display the results were already shown in Sections 1 and 2, so repetition is reduced.

3.1 Motion — Uniform E Field

Special Relativity (SR) modifies classical Newtonian mechanics for energy, momentum, and velocity. In classical physics energy and momentum are separately conserved. In SR energy and momentum are closely related. We take classical kinetic energy to be, $T = p^2/2m$, $p = \text{mv}$. In SR the energy for a material particle, ε, is $\varepsilon^2 = (\text{mc}^2)^2 + (\text{cp})^2$, so that at low momenta $\varepsilon \sim \text{mc}^2 + p^2/2m$, the rest energy plus the non-relativistic, (NR), kinetic energy T. In general, $\varepsilon = \gamma \text{mc}^2$ with $\gamma = 1/\sqrt{1-\beta^2}$. One can approximately think of the "effective mass" gaining a factor of γ due to motion. For momentum $p = \gamma m \beta c$ and again $m \to \gamma m$ is a useful way to think of momentum in SR as the effective mass gaining a factor of γ due to motion, $\beta = v/c$.

Electric and magnetic fields act on charged particles. These fields are used to form beams of particles and to accelerate them to high energies. The simplest fields are uniform over all space. Start with a constant electric field. The units used here will consistently be MKS. They are Coulomb for charge qe, where e is the electronic charge and q is an integer, and Volt/m for the field E. The Lorentz force equation is valid in all inertial frames and it will be used throughout the text. The Lorentz "force" will consistently be interpreted as the time rate change of the SR momentum.

The electric field does work on a charged particle and changes it's energy. The solutions for velocity and position as a function of time follow simply.

$$\beta = \frac{\text{pc}}{\varepsilon}, \quad \epsilon = \sqrt{(\text{pc})^2 + (\text{mc}^2)^2}, \quad p = \gamma\beta\text{mc} = \frac{(\beta\varepsilon)}{c}$$
$$\frac{\text{dp}}{\text{dt}} = \text{qeE}, \quad p = (\text{qeE})t \tag{3.1}$$

The Lorentz force equation in this case is easily solved for and the momentum increases linearly in time. Solving, rather, for the velocity and position as a function of time:

$$\beta = \left(\frac{\left(\frac{p}{mc}\right)}{\sqrt{1 + \left(\frac{p}{mc}\right)^2}} \right) = \frac{at}{\sqrt{1 + (at)^2}}, \quad a = \left(\frac{qeE}{mc} \right)$$
$$z = \left(\frac{c}{a} \right) \left(\sqrt{1 + (at)^2} - 1 \right) \tag{3.2}$$

```
% look at trajectory of a charge in a uniform E field, Eo - analytic
% dP/dt = b = qEo, P = bt, dbeta/dt = a (NR) = qEo/m,  dbeta/dt (SR)
= a/sqrt(a^2+1)
% dp/dt = q*Eo, p = (q*Eo)*t, beta = p/E = at/sqrt((at)^2 + 1),
a = q*Eo/m
```

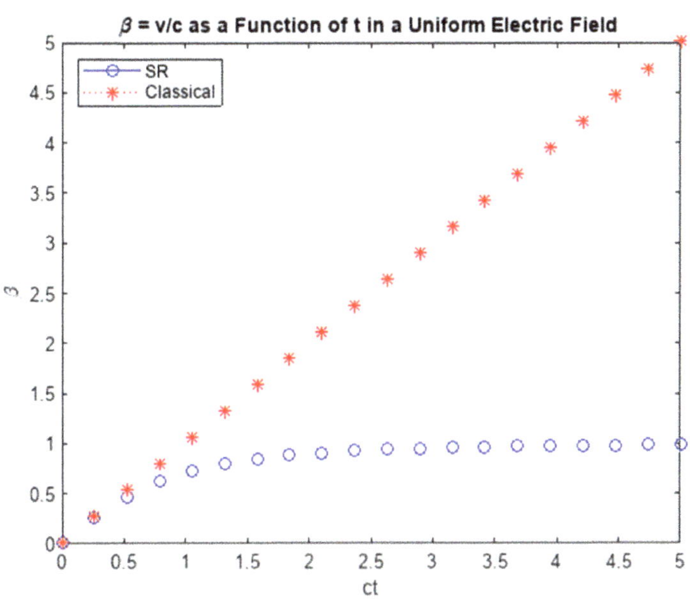

Figure 3.1: Velocity as a function of time for a uniform electric field.

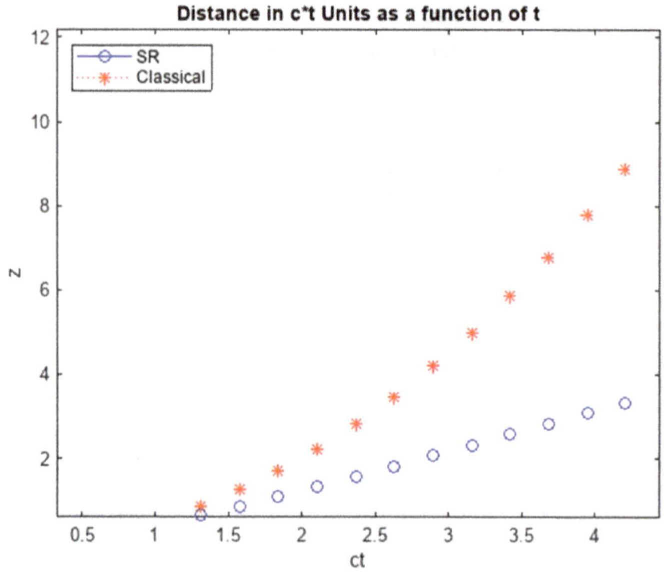

Figure 3.2: Position as a function of time for a uniform electric field.

The "movies" are in constant time steps — so the velocity can be inferred by the spacing of the "frames". In this case the classical velocity increases monotonically, while the SR velocity approaches a constant limit = c.

3.2 Motion — Uniform B Field

The MKS/SI units for the magnetic field are Tesla (T) or Weber/m^2, where a Weber is a measure of the flux of the magnetic field B. A Tesla is 1 N/(Am). The permeability of the vacuum is $\mu_o = 4\pi \times 10^{-7}$(Tm)/A. The Lorentz force is here perpendicular to the velocity, so the magnitude of the momentum is constant and energy is conserved. The direction cosines are α, and they change as the particle bends in the field. The total arc length is the parameter ds = vdt, while the radius of curvature is ρ. The Lorentz force equation is used and the constancy of the momentum p is assumed (no work done by a B field in contrast to the electric field). The charge is always in units of e so that q is a positive or negative integer.

$$\begin{aligned} &\overrightarrow{\mathrm{dp}}/\mathrm{dt} = \mathrm{qe}(\overrightarrow{v} \times \overrightarrow{B}), \quad \overrightarrow{p} = \gamma \mathrm{mv}\overrightarrow{\alpha}, \quad \overrightarrow{\alpha} = \overrightarrow{\mathrm{dx}}/\mathrm{ds} \\ &d\overrightarrow{\alpha}/\mathrm{ds} = (\overrightarrow{\alpha} x \overrightarrow{\alpha}_B)/\rho, \quad \rho = p/(\mathrm{qeB}) \end{aligned} \tag{3.3}$$

In MKS units, e = 0.3 with p in GeV (c = 1 is the convention, but pc has the dimensions of energy), B in T and ρ in m which makes the radius of curvature ρ(m) = 3.33 p(GeV)/B(T). For example, if $p = 100\,\text{GeV}$ and $B = 1$ T, the radius is 333 m.

A useful approximation for small bend angles, ϕ, and magnet length L is that, with motion largely along z, after ds $\sim$ dz $= L$, $\phi = L/\rho$, $y \sim y_o + \alpha_{\text{yo}}L$, $x \sim x_o + \alpha_{\text{xo}}L + \phi L/2$. In terms of a transverse momentum impulse Δp_T, imparted to the particle, $\phi \sim \Delta p_T/\text{p}$, $\Delta p_T \sim \text{eLB} = 0.3\ \text{B(T)L(m)}$ independent of the particle, if $q = 1$. For a 1T field with a $L = 1$ m length, the transverse impulse is 0.3 GeV. The equations are solved exactly below for a uniform field directed along the y axis. The momentum components transverse to the field rotate by an angle ϕ. For position a second integration is made.

$$\begin{aligned} \begin{bmatrix} p_x \\ p_z \end{bmatrix} &= \begin{bmatrix} \cos\phi & \sin\phi \\ -\sin\phi & \cos\phi \end{bmatrix} \begin{bmatrix} p_{\text{ox}} \\ p_{\text{oz}} \end{bmatrix}, \quad \phi = \frac{s}{\rho} \\ \begin{bmatrix} x \\ z \end{bmatrix} &= \begin{bmatrix} x_o \\ z_o \end{bmatrix} + \left(\frac{\rho}{p}\right) \begin{bmatrix} p_z - p_{\text{oz}} \\ p_x - p_{\text{ox}} \end{bmatrix} \end{aligned} \tag{3.4}$$

The B field is along y and the initial position is at $(0, 0, 0)$. The initial momentum is in the (x, y) plane. The B field can be altered to change the radius of curvature, ρ, using a Live "EditField". There is a "movie" of the trajectory which is a helix with $y(t) = y_o + s(\text{py}/p)$, Symbolic solutions, in the special initial case, $\alpha_{\text{xo}} = 0$, of the helical trajectory are displayed.

```
% Charged particle in a constant B field, Bo, along y
% radius r, arc length t
% dp/dt = q(v x B), p constant, radius of curvature r constant
% initially moving along z axis, direction cosines a, rotate in
(x,z) plane
% numeric plots
% Take a step of arc length s in a uniform magnetic field of
magnitude Bo
% along the y axis, charge qe, incident position,xo,yo, momentum
components
% pox, poy poz - exact solutions. Movie made
    ylim([0 60])
    zlim([-8 0])
    hold('on')
end
%
plot3(xxx , yyy, zzz,'-b')
hold('off')
```

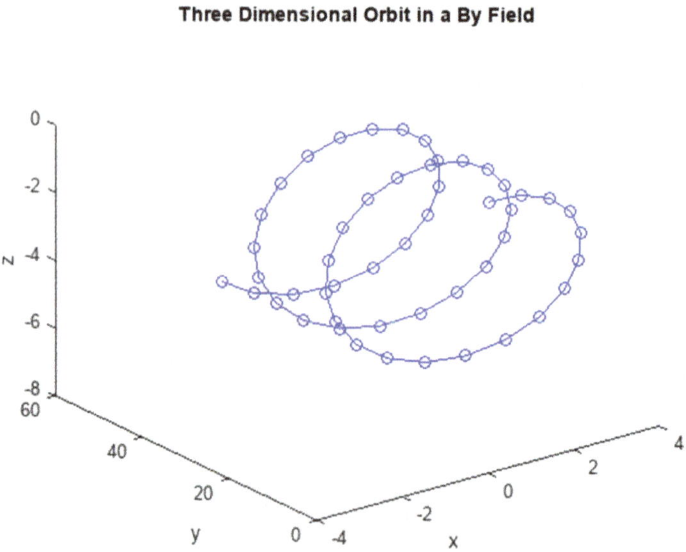

Figure 3.3: Three dimensional trajectory in equal time steps for a uniform magnetic field.

3.3 Non-Relativistic (NR) Cyclotron

The cyclotron is a device with a radiofrequency, r.f., electric field which is in synch with the orbit of the particle beam. There is also a magnetic field which constrains the beam to that circular orbit whose radius increases as the beam energy increases. A crucial element in the operation of the cyclotron is that the rotation frequency is independent of the beam energy, for NR motion, so that the r.f. is always in synch with the beam. The r.f. field is applied in the gap across the "dees" where the circular vacuum chamber is split along a diagonal. The user selects the energy gain for each dee gap traversal, where a large value causes the beam to exit the cyclotron. The number of possible turns is fixed at 80 if the beam does not exit the vacuum tank.

```
% Look at Cyclotron Operation - NR only - p or ion
% work in dimensionless units as possible
% Cyclotron frequency w = qeB/m, r = m*v/qeB = v/w, E = (qeBr)^2/2m
% Relativistic Effects, m -> E, w Decreases and r Increases by gamma
```

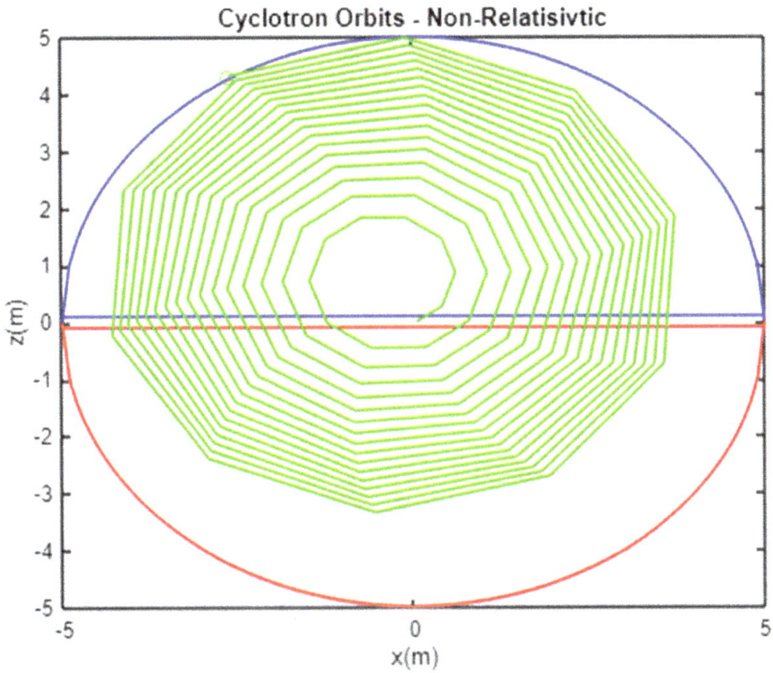

Figure 3.4: Orbit in a cyclotron. The "dees": are in red and blue. The proton orbit, in green, starts at $(0, 0)$ and spirals out as the proton gains energy.

3.4 Solenoids and Fringe Fields

The Lorentz force equation tells us that for ultra-relativistic (UR) charged particles the ratio of the electric to magnetic force is ~E/cB. For a typical 2T magnetic field, an electric field having the same force would be 6 MV/cm, which is quite a large field, difficult to maintain. Therefore, electric fields are usually generated at r.f. frequencies and used almost exclusively for acceleration rather than for steering beams. An exception is the "crossed" E x B fields which are applied for quite low $\beta = v/c$ beam applications.

How might one make a uniform magnetic field? A solenoid carrying a current I with N total turns of length L and radius a has a magnetic field on the center axis of $B = \mu_o(N/L)I(L/2)/\sqrt{a^2 + (L/2)^2}$. A long solenoid has approximately a purely axial field of magnitude $B \sim \mu_o\text{In}$, $n = N/L$. The dimension of μ_o is (Tm/A) and μ_o is 1.26×10^{-6} in those units. For example, with n in turns/cm and I in A, for I = 20,000 A and 1.6 turns/cm the field is 4T. The Compact Muon Solenoid (CMS) experiment at the CERN Large Hadron Collider (LHC) is just such a large 4T solenoid with approximately those parameters. A schematic picture of the CMS solenoid is shown below. The stored energy in the field is 2.7 GJ which requires careful operation. The "fringe field" is also quite large and extends far outside the coil, as will be seen below.

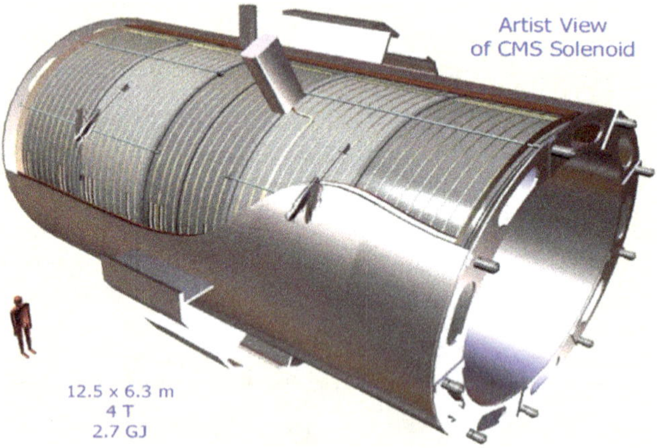

Figure 3.5: Schematic view of the large CMS magnet installed I the CERN LHC.

Construct a "solenoid" out of a series of current loops using the Biot-Savert law to construct a "solenoid" of from 1 to 6 current loops with variable spacing between the loops. The user specifies the number of turns and the turn spacing. Note that the fringe field is unavoidable. This is a feature of Maxwell's equations which force the existence of fringe fields as the main field falls to zero, on a scales dz = d, far from the source currents. The main fringe field is ~ linear in the transverse coordinates which means it behaves approximately like a quadrupole magnet, while the fringe field along the third axis is quadratic and has sextupole behavior. For a main field along the solenoid y axis the z and x fringe fields are approximately:

$$\begin{aligned} &\vec{\nabla} \times \vec{\mathrm{B}} = 0, \quad \partial B_y/\partial_z = \partial B_z/\partial_y \sim \frac{B_o}{d} \Longrightarrow B_z \sim B_o \frac{y}{d} \\ &\vec{\nabla} \bullet \vec{B} = 0 \Longrightarrow B_x = \sim B_o \frac{\mathrm{xy}}{d^2} \end{aligned} \tag{3.5}$$

The user can explore the fringe field as a function of the number of loops, 1, 2, 4 and 6, and the d/a ratio, specifying the spacing between the loops. Try to make a uniform field in the magnet center with a small fringe field. The magnetic "length" of a dipole is normally greater than the physical length of the coils due to the fringe field effect. The quadrupole nature of the fields at the magnet ends may also need to be taken into account in careful magnet models.

```
% Plot B field of a single loop, a pair of current loops, 4, or
6 loops
% numerically - all points using Biot-Savert law
% distance between loops/turns % ihelm = 1,2,3,4 -pairs of
Helmholtz Loops =
```

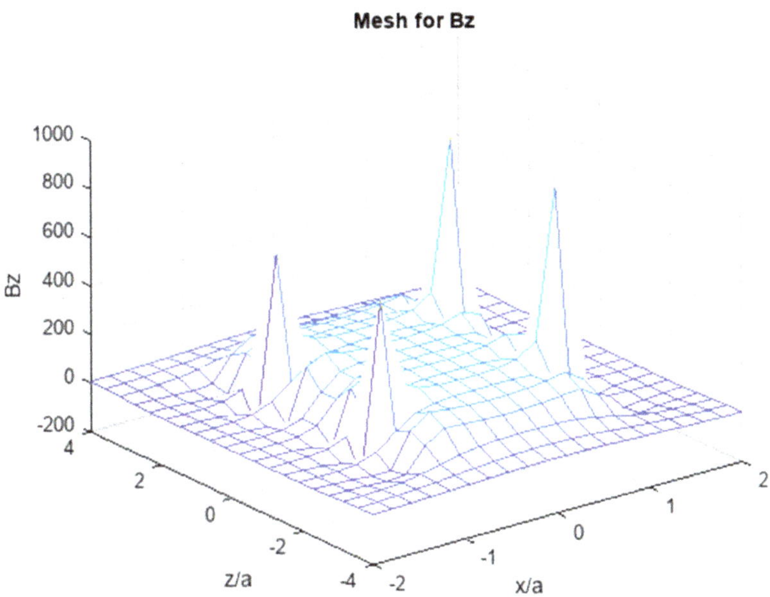

Figure 3.6: Plot of the main field of a configuration of 6 current coils as a function of (x, z).

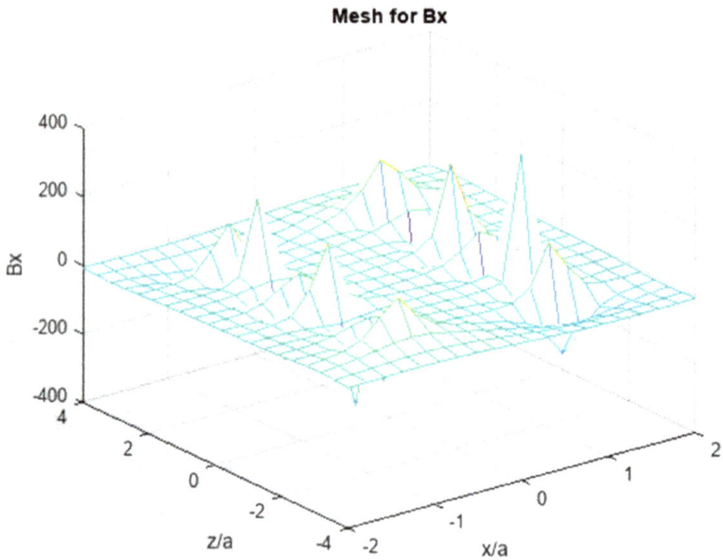

Figure 3.7: Plot of the x fringe field of a configuration of 6 current coils as a function of (x, z).

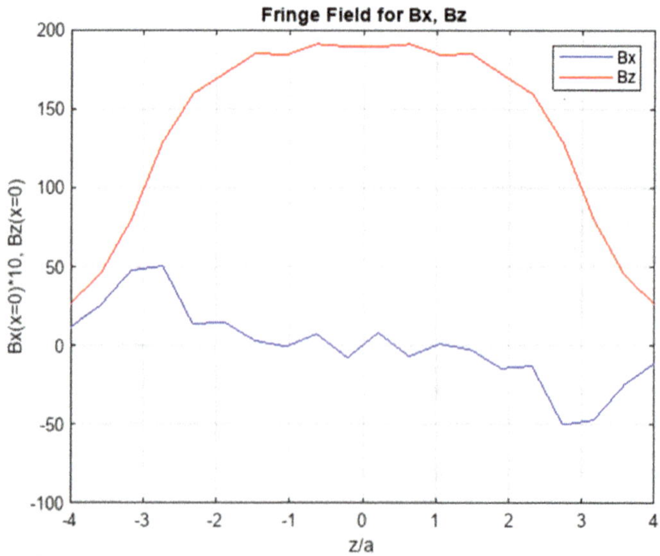

Figure 3.8: Plot of the main field and x fringe field of a configuration of 6 current coils as a function of z at $x = y = 0$.

3.5 "Crossed" E and B Fields

In accelerator produced secondary beams, the momentum is normally defined by dipole magnets and collimators to be within a limited range. There are particles within the acceptance of the beam, all of the same charge, qe, which have different masses and therefore different velocities. Crossed fields, E and B, can select an undeflected velocity and therefore, the mass of a specific particle. Muons have mass (times c^2) 0.113 GeV, charged pions 0.140 GeV. The charged kaon mass is 0.494 GeV and proton mass is 0.938 GeV. One additional method of particle identification is to utilize Cerenkov radiation which is dependent on the velocity of a particle in a medium. This process will be explored later.

The method of velocity selection using uniform "crossed" electric and magnetic fields relies on the fact that there is a specific undeflected velocity, β_o. The deflection angle due to the E × B fields is small, typically, and is normally evaluated approximately. Recall that the electric field is normally weak compared to the magnetic field for UR particles.

$$\begin{aligned} \vec{F}_{\mathrm{E\times B}} &= \mathrm{qe}\left(\vec{E}_o + \vec{\beta} c \times \vec{B}_o\right) \\ \text{if } \beta = \beta_o &= \frac{E_o}{(cB_o)}, \quad \vec{F}_{\mathrm{E\times B}} = 0 \end{aligned} \tag{3.6}$$

If the electric and magnetic fields are perpendicular to each other (directed along x and y) and also perpendicular to the particle velocity (along z), a particular velocity has no force acting on it. For the angular deflection L is the length over which the E and B fields exist. Other velocities will be deflected by the crossed fields. For small deflections, there is a simple approximate solution where the time in the fields is taken to be the undeflected time and a small angle approximation is used.

$$\begin{aligned} F_y &\sim \frac{\mathrm{dp}_y}{\mathrm{dt}}, \quad \mathrm{dt} \sim \mathrm{L}/(\beta c) \\ d\theta_y &\sim \frac{(dp_y)}{p_o} \sim F_y \frac{L}{(\beta c p_o)} \sim (\mathrm{qeE}_o)\frac{L}{(cp_o)}\left(\frac{1}{\beta} - \frac{1}{\beta_o}\right) \end{aligned} \tag{3.7}$$

For example, at the KEK, a particle physics laboratory in Japan, in the electrostatically separated Kaon beam, Eo ~ 5.5 MV/m, Bo ~ 0.02 T so that $\beta_o \sim 0.916$. For a beam with a central momentum, $p_o = 1.1$ GeV/c, pions have a β of 0.992, while kaons have a value of 0.912, and are undeflected. Over a length L of 0.4 m, the pions get a vertical angle kick of ~0.18 mrad. At a momentum collimator (mass slit) 15 m downstream of the fields, the pions are deflected by only 2.7 mm and absorbed in the collimator. Note that the chosen low beam momentum and low magnetic field are the result of the relative weakness of the electric force compared to the magnetic force, as mentioned previously.

For the exercise, a full treatment of crossed fields is possible and MATLAB has the symbolic and numerical tools to make short work of the problem. First do the symbolic setup of the differential equations in this case where momentum is not conserved. That leads to terms proportional to the variation of γ. For a particle in crossed E and B fields, where E is along y, B is along x and the beamline is z with $s \sim z \sim$ ct, the Lorentz force equations $\vec{F} = q(\vec{E} + \vec{v} \times \vec{B})$ are shown below. It is an interesting fact that these equations are valid in any inertial reference frame as long as all the measurements are made in the same frame,

$$\begin{aligned}
&\beta_x \frac{\mathrm{d}}{\mathrm{d}s}\gamma + \gamma\frac{\mathrm{d}}{\mathrm{d}s}\beta_x = 0 \\
&\beta_y \frac{\mathrm{d}}{\mathrm{d}s}\gamma + \gamma\frac{\mathrm{d}}{\mathrm{d}s}\beta_y - E + B\beta_z = 0 \\
&\beta_z \frac{\mathrm{d}}{\mathrm{d}s}\gamma + \gamma\frac{\mathrm{d}}{\mathrm{d}s}\beta_z - B\beta_x = 0 \\
&E = \frac{\mathrm{qeE}_o}{\mathrm{mc}^2}, \quad B = \frac{\mathrm{qeB}_o}{\mathrm{mc}}, \quad \beta_o = \frac{E_o}{\mathrm{cB}_o}
\end{aligned} \tag{3.8}$$

One might be tempted to use the powerful MATLAB tools to solve the equations exactly. The momentum is not constant in an electric field which makes the solution rather complex, since γ is now time dependent. The results are cleaned up using the utilities "subs" and "simplify".

```
% Lorentz force for E x B, E along y, B along x, beam incident
along z, dp/dt
% dp/dt has d(gamma)/dt and d(beta/dt). p not constant
```

$$\texttt{sdbx} = -\frac{\mathrm{bx}(t)(\mathrm{Eby}(t) + B\mathrm{bx}(t)\mathrm{bz}(t) - B\mathrm{by}(t)\mathrm{bz}(t))}{g}$$

$$\texttt{sdby} = -\frac{B\mathrm{bz}(t) - \mathrm{E} + \mathrm{Eby}(t)^2 - B\mathrm{by}(t)^2\mathrm{bz}(t) + B\mathrm{bx}(t)\mathrm{by}(t)\mathrm{bz}(t)}{g}$$

$$\texttt{sdbz} = \frac{B\mathrm{bx}(t) - \mathrm{Eby}(t)\mathrm{bz}(t) - B\mathrm{bx}(t)\mathrm{bz}(t)^2 + B\mathrm{by}(t)\mathrm{bz}(t)^2}{g}$$

One should note that the equations can be approximated simply. This is the "spherical cow" approach of simplifying the problem. The x and y velocities are taken to be small, so that 0 is a good first approximation for their time rate of change. The derivative of the z velocity is also small so it can be approximated as the initial value. The y velocity change, along the electric field direction is approximately $d\beta_y/\mathrm{ds} \sim \beta_z B - E \sim B - E$ which illustrates the drift velocity, since $\mathrm{E/cB} = \beta_o$. One lesson is that it is crucial not to get lost in the algebra. First, make approximations and see how the solutions to the simplified equations behave. Indeed, the y angular kick was already estimated above to be small.

The full solutions can now be explored numerically for a typical application — the KEK kaon beam. The MATLAB differential equation numeric solver, "ode45" is the utility of choice. One finds that the approximate solution is quite adequate to represent the full blown solution, which is a useful conclusion. One should start simply and add complexity only when it is warranted or needed.

```
% crossed E and B, Eo along y, Bo along x, beam initially (0,0,0)
% initial velocity along z, dp/dt = qe* (E + v x B) - NR
% dp/dt in SR case, p = g*b*m*c, MKS units
Ey = 5.5e6; % KEK electric field in MV/m
po = 1.1; % K beam momentum K beta = 0.917, pion beta = 0.997
% B field in T for K selecton, this
```

```
bK = 0.9123
betapi = 0.9920
Bx = 0.0201
```

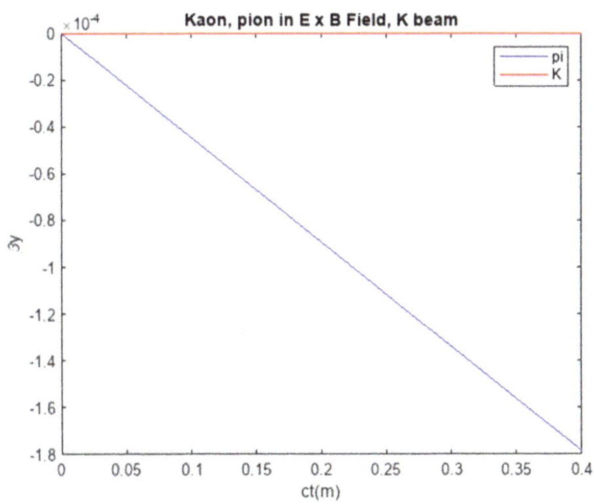

Figure 3.9: Plot of the y velocity of pions and kaons as a function of distance in crossed electric and magnetic fields.

3.6 Thomson Scattering, Larmor Radiation

The Thomson cross section describes the scattering of low energy photons off a localized charge, taken here to be a single free electron. The angular distribution is dipole. The cross section for this process is smaller than the geometric cross section by a factor $\sim\alpha^2$ because it occurs by way of the interaction of the photon with the electron via the electromagnetic field. It can be visualized as the photon being absorbed by the electron which subsequently decays into an electron and a photon. At very low energies the cross section for an atom is just Z times the free electron cross section because the photon scatters incoherently off all the electrons in the atom. The interaction is assumed to be elastic; no additional particles are created.

The cross section is normalized to the incoming photon energy flux using the Poynting vector, $\overrightarrow{S}$, which is the cross product of the electric, $\overrightarrow{E}$, and magnetic fields, $\overrightarrow{B}$, of the photon and has dimensions of power, P, per unit area. For an incident plane wave with amplitude E_o of the electric field the time averaged Poynting vector is $\varepsilon_o c E_o^2/2$. The incident flux is not intrinsic to the physics of the process and so it is factored out leaving the Thomson cross section, σ_T. The Compton wavelength of the electron is $\lambda = h/(m_e)$, the fine structure constant is $e^2/\hbar c \sim 1/137$ and r_e is the classical electron radius.

$$\begin{aligned} d\sigma_T/d\Omega &= [d\langle P\rangle/d\Omega]\langle |\overrightarrow{S}|\rangle, \quad \overrightarrow{S} = \overrightarrow{E} \times \overrightarrow{B} \\ &= (\alpha\lambda_e/2\pi)^2 \sin^2\theta \\ \sigma_T &= (8\pi/3)(\alpha\lambda_e/2\pi)^2 = 0.66\ b, \quad r_e = (\alpha\lambda_e/2\pi) \end{aligned} \tag{3.9}$$

NR radiation by an accelerated charge has an angular distribution in lowest multipole order which is also dipole, with emission preferentially perpendicular to the acceleration (Larmor). In a magnetic field, the Lorentz force equation implies a rotation of the charged particle which has the instantaneous circular frequency, $\omega_o = \mathrm{qeB}/m$. The radiated power (NR) is for a dipole of moment p_o oscillating at a frequency ω_o and scales as the fourth power of ω_o

$$\begin{aligned} d\langle P\rangle/d\Omega &= \omega_o^4 p_o^2 \sin^2\theta/(8\pi\varepsilon_o c^3) \\ \langle P\rangle &= \left(\omega_o^4 p_o^2\right)/(12\pi\varepsilon_o c^3) \end{aligned} \tag{3.10}$$

In the case of extremely relativistic beams, for example from electron accelerators, the radiation pattern changes dramatically. To begin by only looking at the transformation of photons, the Lorentz transformation defines the relationship between polar angles in different reference frames independent of any dynamics. This is purely kinematics. The forward emission of photons is called the "search light" effect. It is a general effect due to SR. The transformation for photon energy and momentum implies a simple relationship between photon angles for an isotropic distribution in one frame, the $*$ frame, transformed to a distribution in a frame moving with velocity βc. The relationships are: $p = \gamma p^*(1 + \beta \cos\theta^*)$, and $p\cos\theta = \gamma p^*(\cos\theta^* + \beta p^*)$. For reference frames in extreme relative motion an isotropic photon angular distribution becomes sharply peaked in the direction of motion of the reference frame. The maximum angle is found using the utility "max" and the value is $\sim 1/\gamma$. If $\gamma = 1$ the distribution is isotropic. The specific case of synchrotron radiation in circular motion is deferred until the discussion of relativistic effects. The user can choose the γ and see the resulting locus of angles over the full angular range in the $*$ reference frame.

```
% Lorentz trans, pz and p
% p = g*(p* + b*costst) - energy
% pcost = g*(p*costst + b*P*) - pz
% Jacobian of trans of dP/domega* = 1/4*pi, isotropic rest frame
```

$$\frac{1}{g^2(b\cos t - 1)^2}$$

```
% polar plot
polarplot(t,dsig)
% search light effect, max theta ~ 1/gam
```

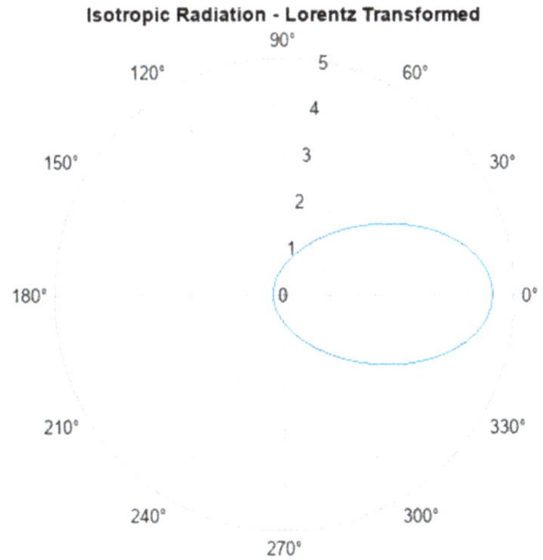

Figure 3.10: Polar plot of the angular distribution of Thompson photon scattering with a user chosen photon energy.

3.7 NR Dipole Radiation

Non-relativistic, NR, radiation of an accelerated dipole — an antenna — has a characteristic dipole $\sin^2(\theta)$ shape. The angle is that between the wave vector, k, and the dipole axis. The dipole is driven harmonically and is the accelerated charge distribution. This shape will alter dramatically for relativistic acceleration, as will be seen later in the text. The form of NR dipole radiation is simple in the limit that r and 1/k, the inverse wave vector, are much greater than the size of the radiating dipole, which is called the "far zone", The radiated energy falls off as the inverse of the radius, which means that the power crossing a sphere of radius r is independent of r. That is what is meant by radiation. Static electric fields fall off typically much more rapidly with radius. The dipole moment oscillates as $p_o(t) = p_o \cos(\omega t)$. The electric field is: E and the Poynting vector is $\vec{S}$.

$$
\begin{aligned}
E &= (-\omega^2/(4\pi\varepsilon_o c^2 r))p_o \cos(\omega t - r/c)(\sin\theta/\theta) \\
\vec{S} &= (1/4\pi\varepsilon_o)^2(1/c^5 r^2 \mu_o)\omega^4 p_o^2 \cos^2(\omega(t - r/c))\sin^2\theta(\vec{r}/r)
\end{aligned}
\tag{3.11}
$$

The power, P, into a solid angle $d\Omega$ is shown below. The time averaged power integrated over all angles is $\langle P\rangle\omega^4 p_o^2/(12\pi\varepsilon_o c^3)$:

$$
\mathrm{dP}/d\Omega \sim (\mathrm{k}\sin^2\theta/r)\sqrt{1 + (1/\mathrm{kr})^2(\cos(k(\mathrm{ct} - r) + \tan^{-1}(\mathrm{kr})))}
\tag{3.12}
$$

A "movie" is made of the dipole radiation pattern where the dipole oscillation itself is also indicated. The value of the wave vector can be varied in order to see how the dipole pattern depends on the wave vector magnitude. The trough in the power seen in the movie, is due to the dipole sin factor.

```
% Dipole radiation - contours of power. Far zone
% Dipole radiation when r and 1/k are much larger than the dipole
size
```

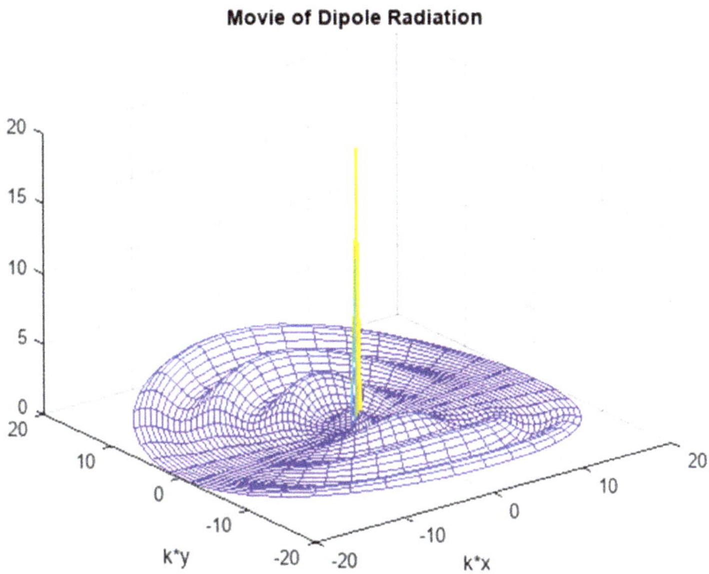

Figure 3.11: Dipole radiation as a dipole (vertical yellow) oscillates in time and emits outgoing radiation.

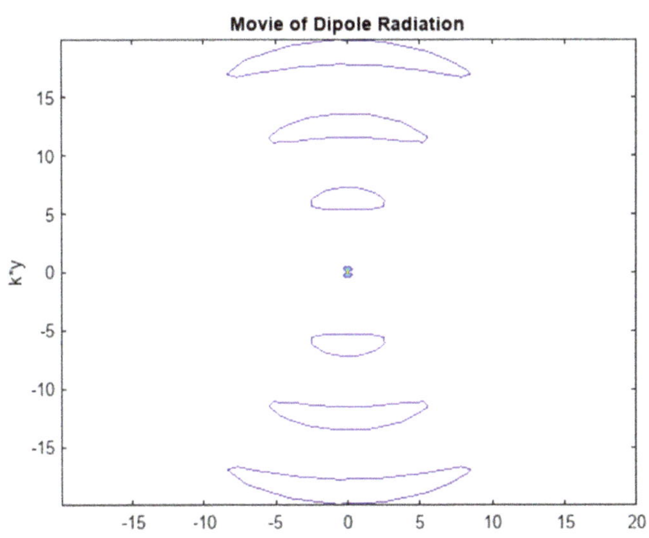

Figure 3.12: Dipole radiation contours as a dipole oscillates in time and emits outgoing radiation.

3.8 Cerenkov Cone

The use of Cerenkov radiation allows for the extension of particle identification (ID) to higher values of velocity than the use of ionization or time of flight. A charged particle emits radiation if it exceeds the velocity of light in some medium, c/n. The medium is here defined by an index of refraction n, assumed to be real. The Cherenkov angle, θ_c, is the angle of the photon with respect to the particle direction. Light is not emitted below a threshold velocity, $\beta_{\rm th} = 1/n$. These properties are made visible using the code below. The user selects the velocity of the particle in the medium, either below or above threshold. When above threshold the outgoing spherical waves align into a "Mach cone" whose angle increases as the velocity increases. In this case, $\cos\theta = v_s/v$ where v_s is the velocity of signal propagation, c/n. A "movie" is made of the outgoing waves for 6 distinct points of emission labelled by the green dots on the x axis. The forward waves are blue, backward red.

Detectors utilizing the angle of the Cerenkov radiation of particles of differing masses in a momentum selected beam can directly detect the light at a selected angle, and hence the mass. A cruder measurement with no angular information simply detects the presence of absence of light, which are called threshold detectors since all particles with velocity greater than the threshold give light.

```
% make for 6 emission times and points
% sample at 10 times
```

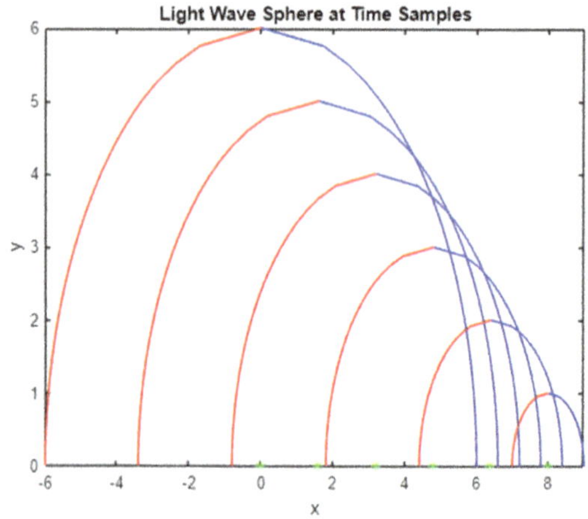

Figure 3.13: Plot of locus of forward (blue) and backward (red) emission when velocity of emitter exceeds the velocity of light in the medium.

3.9 Image Charge and Beam Position

The first exercises for electromagnetism largely concerned themselves with dynamics, motion in defined E and B fields with possible consequent photon emission. Now the static solutions for E and B fields due to specific sources, either charges or currents, is examined.

Consider a beam of charged particles. It can be considered a current source which induces image charge on a detector of the beam position. First find $d\varepsilon/d\beta$. This quantity for an ultra-relativistic (UR) beam is very large which means physically that it requires a lot of energy to change the velocity when the velocity is ~1.

```
e = m*c*c/sqrt(1-beta*beta);
dedbeta = diff(e,beta); dedbeta = subs(dedbeta,sqrt(1-beta*beta),
1/gamma)
```

`dedbeta =` $\beta c^2\gamma^3 m$

The number of charges/length in the beam with bunch length L is n = N/L where N is the total number in the bunch. The UR impedance Z_b is very large because it is very hard to change the beam velocity, and is inversely proportional to the beam current, when a voltage, V, is applied for UR beams. Therefore, the beam can be considered to be an ideal current source in most situations.

$$I_b = \mathrm{qen}\,\beta c, \quad Z_b = \left(\frac{\mathrm{dV}}{\mathrm{dI}_b}\right)$$

$$\varepsilon = \gamma \mathrm{mc}^2, \quad \frac{(d\varepsilon)}{d\beta} = \mathrm{mc}^2\beta\gamma^3, \quad d\varepsilon = \mathrm{qed}V \tag{3.13}$$

$$Z_b = \frac{\left(\mathrm{mc}^2\beta^2\gamma^3\right)}{\mathrm{qeI}_b}$$

To find the induced current for a cylindrical conducting beam pipe, I, one can use the method of image charges for a cylindrical geometry, which is taken to be only 2 dimensional, (x, y), since the z range of a UR beam is small and Ez for a passing beam has 0 time integral. The solution is obtained as a function of the (x, y) location inside the beam pipe. For a charge qe with a radius r inside the pipe of radius a the image charge is along the radius to the charge with image charge $q_i = -\mathrm{qe}(\mathrm{a}/\mathrm{r})$ at location $r_i = a^2/\mathrm{r}$. The image plus beam charge induces a linear current density j on the wall at (a, Φ) due to the beam charge at (r, ϕ). Integrating j over a small range of azimuth α of the assumed pickup electrode, the induced current is approximately the full beam current reduced by the azimuthal fraction subtended by the pickup.

$$j(a, \Phi) = -\left(\frac{I_b}{2\pi a}\right) \frac{[(a^2 - r^2)]}{[a^2 + r^2 - 2\,\mathrm{racos}(\Phi - \phi)]}$$

$$I = a \int_{-\frac{\alpha}{2}}^{\frac{\alpha}{2}} j(a, \Phi) d\Phi \sim I_b \left(\frac{\alpha}{2\pi}\right) \tag{3.14}$$

The "Sliders" change the radial and azimuthal location of the point charge inside the beam pipe. The contours of the image charge are plotted. An off axis beam has a ϕ asymmetric induced charge density while a beam charge at $(0, 0)$ induces equal charge at the 2 opposite sides of the beam pipe. The location of the beam can be changed in order to see how the induced current density changes. The position of the beam can be inferred by measuring the induced charge at 2 locations at the beam pipe on opposite sides of a diagonal.

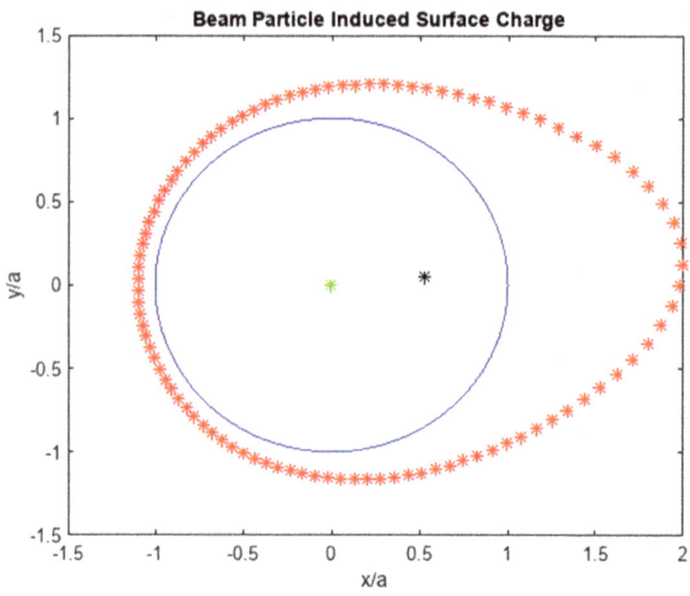

Figure 3.14: Contour of image charge (red dots) for an off center beam (black dot) inside a cylindrical conductor.

3.10 Fast Fourier Transform — FFT

The Fast Fourier Transform (FFT) is applied to a sequence of N time samples. The FFT is expanded in a series of N harmonic terms. This series can be inverted and for real functions the expansion is in terms of N cos and sin functions, analogous to the Fourier series coefficients. These operations are available in MATLAB using the utilities "fft", "ifft" to find the FFT and the inverse FFT of a series of data points, and the utility "abs" which is used to find the power as a function of frequency. The FFT operation is on a discrete series of data points, not a function. In this text only the power is explored and not the phase information that is also available using the "fft" utility.

"FFT_Sunspots" is supplied to introduce the MATLAB FFT utilities. In fact, MATLAB supplies 288 years of data on sunspot activity. The data is displayed and the FFT is found. Then the FFT power is found as a function of frequency or period and the maximum power is computed. The sunspot cycle has a period of 11.08 years. The plots of sunspot activity in both time and frequency are displayed. The two descriptions are, indeed, complementary. The user can look at the code with the "comment" lines in order to become familiar with the "fft" utility. Alternatively, there is MATLAB documentation available using the "Search Documentation" window.

In what follows, electrostatic problems without interior sources will be explored. Then the effect of interior sources will be examined using the FFT technique.

```
load sunspot.dat; % MATLAB data on sunspot activity per year
```

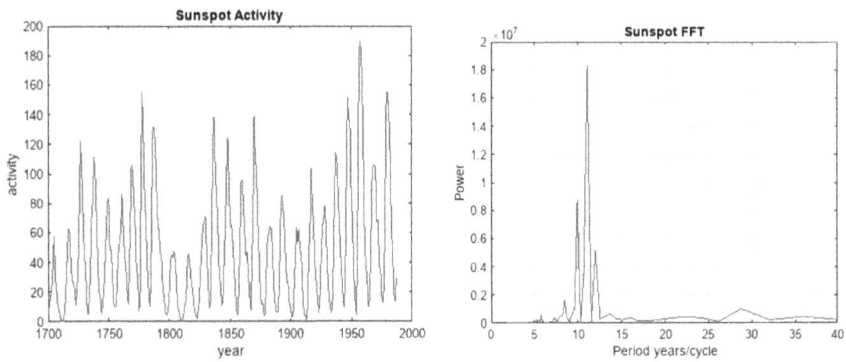

Figure 3.15: Annual sunspot data (left) and FFT power spectrum (right).

```
% sun FFT
[mxP, imx] =  max(SP); % max power
Stau(imx)
```

```
ans = 11.0769
```

3.11 Laplace Equation — Numerical Boundary Values

In the absence of charges as sources of the electric field, the boundary values of the potentials uniquely determine the fields. The Laplace equation, $\nabla^2 V = 0$, for the electric potential inside the boundaries, approximated on a finite granularity grid of points, is:

$$\nabla^2 V = 0 \sim 4V(i,j) - V(i,j+1) - V(i,j-1) - V(i+1,j) - V(i-1,j) \tag{3.15}$$

The solution is determined on a 2 dimensional grid of (i,j) points. In the particular code example shown here, any function of x or y on the Cartesian boundaries can be defined along with constants a and b. The user is encouraged to explore the results of changing the "edit field" for both constants and functional dependence. The number of grid points, ng, could also be changed if desired. The number of iterations could also be changed if one wished to see how fast the solution converges. In addition, given the voltage, the electric field could be derived using the "gradient" utility for additional plotting as desired.

```
% Solve Laplace Eq. - sourceless using Gauss- Seidel,
% Cartesian, Boundary Voltages are functions
% Solve Finite Difference Eq: 4Vi,j = Vi,j+1 + Vi,j-1 + Vi+1,j +
Vi-1,j
% grid is 1 x 1 in x x y
ng = 30; % # of grid points = ng x ng
del = 1.0 ./(ng-1.0); % grid spacing
xx = (0:ng-1) .*del;
yy = (0:ng-1) .*del;
```

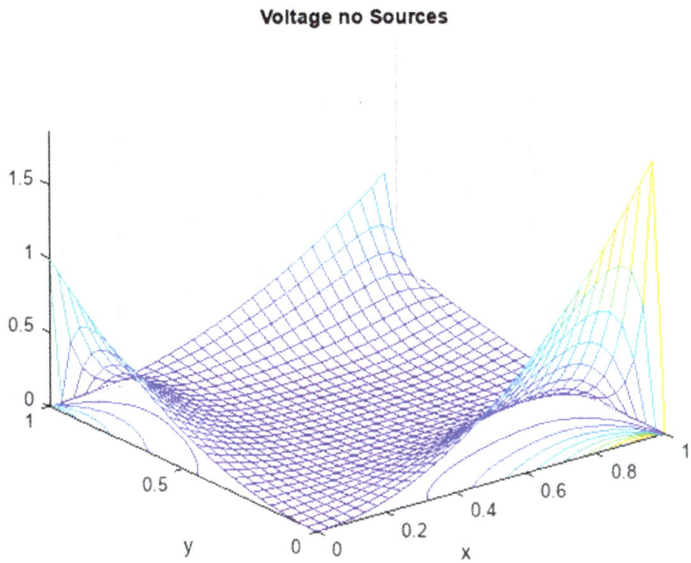

Figure 3.16: Contour of the interior voltage for user defined voltages on the rectangular boundaries.

3.12 Laplace Equation and Complex Variables

Solutions to the Laplace equation in 2 dimensions are well matched to the use of complex variables. Any analytic function of the complex variable z will satisfy the Laplace equation for both the real and complex parts separately. If the result for a simple case is known, more complex situations can be treated by making specific mapping transformations of the solution. Several examples are known and 5 distinct transformations are set up and can be chosen by the user.

```
% Complex psi - Modulus of psi is the Potential V, grad(V) =
Electric Field
% Initialize parameters, x,y grid, and potential psi
ng = 40; % # of grid points
x = linspace(-2, 2, ng);
y = linspace(-2, 2, ng);
% Electrostatic Examples
% 1 =  point charge, 2 = point charge  and plane, 3 = 2 circles
halves, interior
% 4 = 2 coaxial cylinders, interior. 5 = 2 line charges on the y=0
axis
```

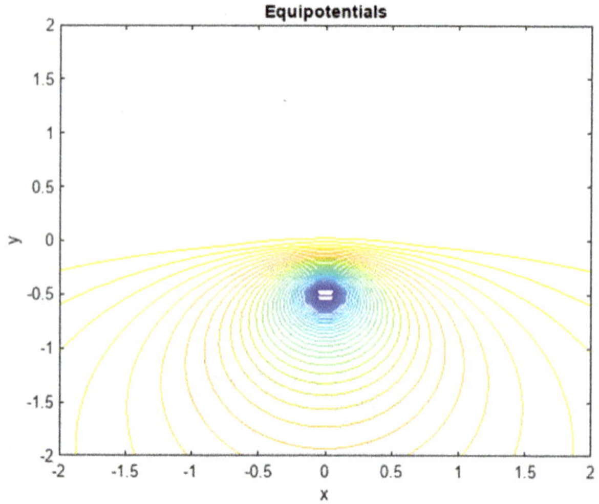

Figure 3.17: Voltage contours in the lower half plane for a point charge and a plane at $y = 0$.

```
[vx,vy]=gradient(psi,x(2)-x(1),y(2)-y(1));
quiver(x,y,vx,vy)
title('Electric Field')
xlabel('x')
ylabel('y')
```

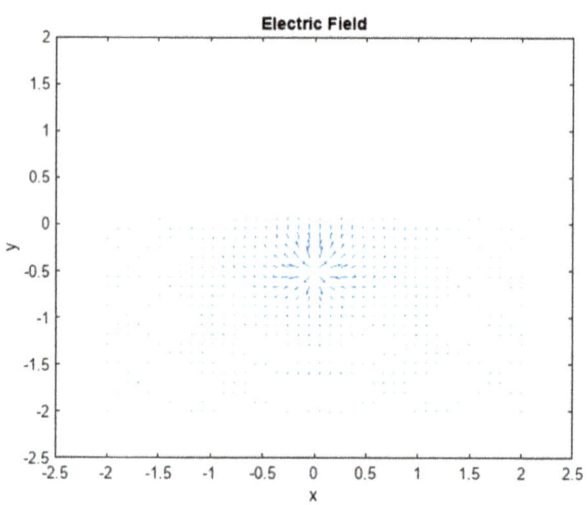

Figure 3.18: Electric field for the potential shown in Figure 3.17.

3.13 Poisson Equation and FFT

The Poisson equation, which incorporates sources, can be solved in general numerically on a grid. Specifically, FFT techniques will be used now for a Cartesian geometry. The sources appear as a charge density, ρ where in 2 dimensions with grid spacing, d, the Poisson equation, $\nabla^2 V = \rho/\varepsilon_o$, approximated on a grid is:

$$\begin{aligned} 4V(i,j) &= V(i,j+1) + V(i,j-1) + V(i+1,j) + V(i-1,j) \\ &\quad + \rho(i,j)d^2/\varepsilon_o \end{aligned} \tag{3.16}$$

The Poisson equation could be solved on a grid as an option but using the FFT utilities which are also available is done here as an example, The utilities used are "fft2" and "ifft2" For a grid $N \times N$ the FFT solution is:

$$V_{i,j}^{\mathrm{FT}} = \rho_{i,j}^{\mathrm{FT}} d^2/[2(\cos(x_i) + \cos(y_j) - 4)]\varepsilon_o, \quad x_i = 2\pi i/N, \quad y_j = 2\pi j/N \tag{3.17}$$

There are charged sources and the solution is harmonic, the cos terms, so that the solution vanishes on the boundaries. That is an important constraint and unless one restricts oneself to be far from the boundaries, the found solutions may diverge from the correct solutions. The user can select the number of charged rectangles and their locations are fixed by supplying the 4 corners. The worked example represents a parallel plate capacitor.

```
% Solve static Poisson Eq. using FFT for Periodic BC, Cartesian
% solve Poisson Eq numerically - 2-D Cartesian Only
% solve Finite Difference Eq: 4Vi,j = Vi,j+1 + Vi,j-1 + Vi+1,j +
Vi-1,j - rhoi,j*del*del
% look at a parallel plate capacitor - with fringe fields
```

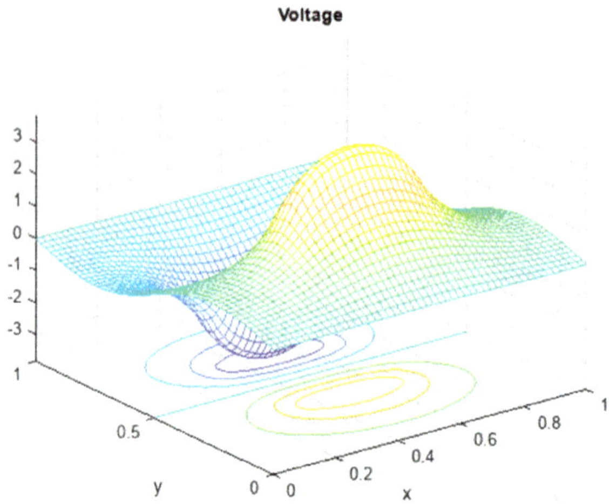

Figure 3.19: Potential for the case of 2 thick plates with opposite voltages.

```
[Ex,Ey] = gradient(real(V'))
```

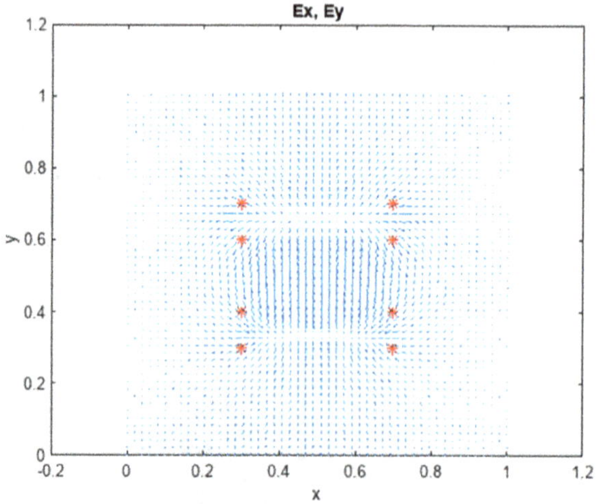

Figure 3.20: Electric field for the case of 2 thick plates with opposite voltages. Note the concentration of field between the plates and the fringe fields outside the plate boundaries.

```
% The field is concentrated between the "capacitor" plates, but
there is a
% substantial "fringe field".
```

3.14 Magnetic Shielding

In many situations it is desirable to reduce stray magnetic field from some sensitive piece of apparatus. In order to accomplish that, a typical solution is to surround the apparatus with a material with a high magnetic permeability, μ. The simple model examined here is a hollow sphere immersed in a uniform external magnetic field B_o along the z axis. The shell has inner radius $a = 1$ and outer radius b made of a material with μ value $\gg$1. In that approximation the reduction in field inside the shell is approximately $9/[2\mu(1-(a/b)^3)]$. which scales approximately as $1/\mu$. If b is much larger than a (thick shielding) the shielding is most effective. In the case of conductors the free charges move so as to cancel any electric field inside the conductor. For dielectrics complete cancellation is not achieved nor is it the case with diamagnetic materials.

The user chooses the thickness of the shell. and the μ value of the material of the shell. The resulting effect on the inner volume is quite large for thick shells made from materials with large μ values.

```
% External B field along z Axis. Spherical coordinates;
% Magnetic potential = -Bo*z Far From the Sphere, Take Bo = 1
mu = 43; % Magnetic Permeability u > 1
% Field Inside the Sphere is Reduced by a Factor ~ 9/2*u if b^3 >> 1
% exact solutions - Jackson, a = 1 and Bo = 1, Mag Potential
% field reduction inside, evaluate magnetic potential
```

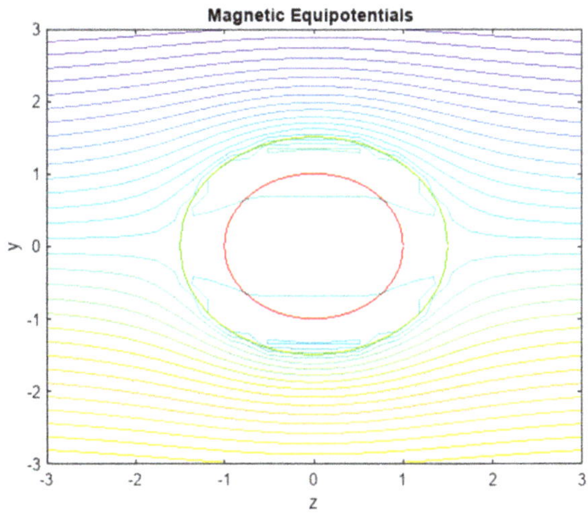

Figure 3.21: Magnetic potential contours for a coaxial cylinder with a high magnetic permability immersed in a uniform magnetic field.

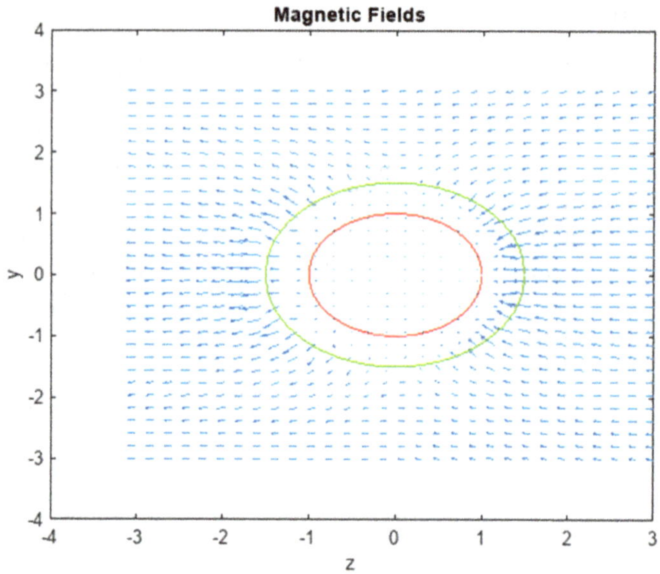

Figure 3.22: Magnetic field for a coaxial cylinder with a high magnetic permability immersed in a uniform magnetic field.

3.15 Dielectric Sphere in an E Field

As an example of a boundary value problem with dielectrics consider a dielectric sphere, dielectric constant $\varepsilon, D = \varepsilon E$, of radius r. There are charges within the dielectric that rearrange themselves in the presence of an electric field such that at an interface between 2 media, the tangential E field is continuous, and the normal D field is continuous. There is an induced surface charge density, σ. There is an external field E_o far from the sphere which points along the z direction and in which the sphere is immersed. Matching the boundary conditions for E and D, the potential V inside and outside the sphere is:

$$
\begin{aligned}
&A = E_o \,\mathrm{rcos}(\theta) = E_o z \\
&V_{\text{in}} \;= -3A/(\varepsilon + 2) \\
&V_{\text{out}} \;= -A[1 - (\varepsilon - 1)]/\left[(\varepsilon + 2)r^3\right] \\
&\sigma/E_o \cos(\theta) = (3/4\pi)(\varepsilon - 1)/(\varepsilon + 2)
\end{aligned}
\tag{3.18}
$$

The user can vary the dielectric constant. For $\varepsilon \sim 1$ there is little disturbance of the external field, while for large values the field inside the sphere is largely excluded, similar to the case where a conducting sphere, with fully mobile charge carriers is immersed in an electric field. Both the potential and field are displayed. It is evident that the exterior field tends to flow around the surface of the sphere, particularly for large values of the dielectric constant.

```
% Fields for a dielecric sphere immersed in a uniform Ez field
% Electric Potential = -Eo*z Far From the Dielectric, Eo = 1
k = 5.2 ; % Dielectric Constant > 1
% exact solution
%  outside dielectric
            phi(i,j) = - A .*(1.0 - (k - 1) ./((k + 2) .*(r .^3)));
% inside
            phi(i,j) = - (A .*3)./(k + 2);
```

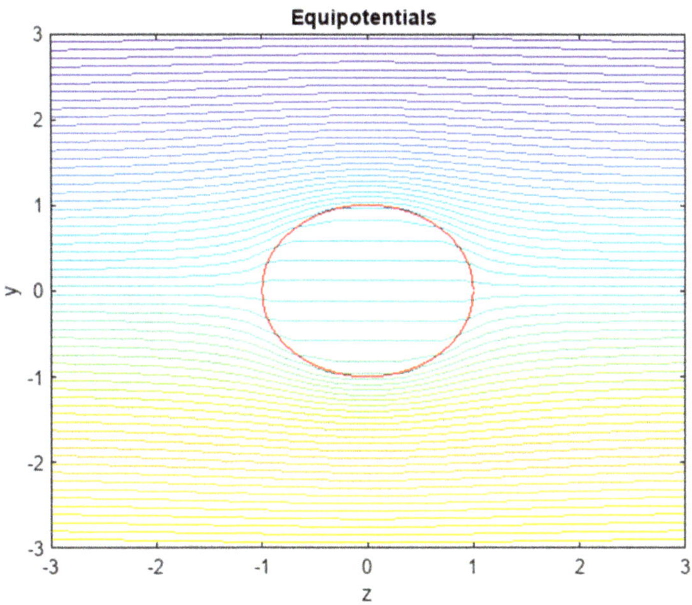

Figure 3.23: Electric potential contours for a dielectric sphere with a high dielectric constant immersed in a uniform electric field.

```
%
[ez,ey] = gradient(phi');
quiver(zz,yy,-ez,ey);
title('Electric Fields ')
xlabel('z')
ylabel('y')
hold on
plot(x1,y1,'r-',x1,-y1,'r-')
hold off
```

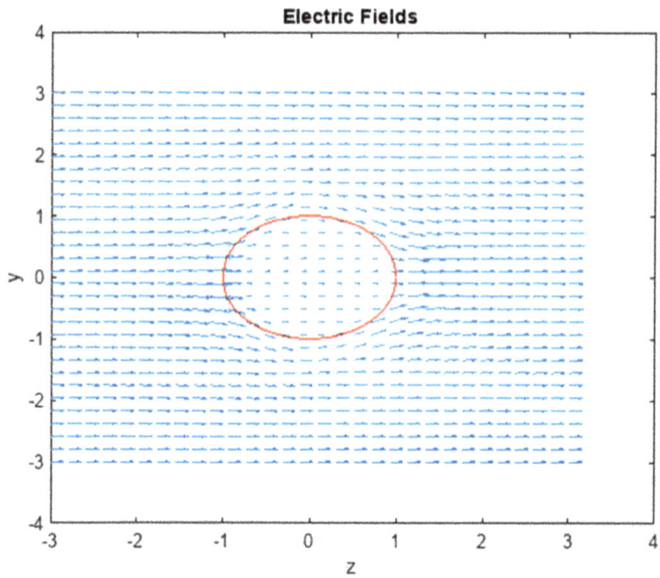

Figure 3.24: Electric field for a dielectric sphere with a high dielectric constant immersed in a uniform electric field.

Chapter 4

Gases and Fluid Flow

"Classifications like "optics" or "thermodynamics" are just straitjackets, preventing physicists from seeing countless intersections." Ted Chiang

"Big whorls have little whorls, which feed on their velocity, and little whorls have lesser whorls, and so on to viscosity." Lewis Richardson

In this Section there will first be a look at classical and quantum statistical mechanics. The classical Boltzmann distribution will then be applied to gases and other dilute systems. In parallel, in order to gather the math in a central place, the quantum statistics of bosons and fermions will be explored at the same time. The physics is in the weight factors, which differ profoundly for classical Bose-Einstein and Fermi-Dirac statistics. The phase space factors have a more general applicability.

4.1 Energy Weight Factors — Maxwell-Boltzmann, Fermi-Dirac, Bose-Einstein

There are fundamentally only 2 distributions. For spin 1/2 fundamental particles, for example the electron, the particles are constrained to not occupy the same quantum state (the Fermi Exclusion Principle). For integral spin particles, for example photons, no such exclusion operates and any number of particles can occupy the same quantum state. The 3 energy weight factors for Maxwell-Boltzmann, Fermi-Dirac, and Bose-Einstein, with $U = (E - \mu)/\mathrm{kT}$ are:

$$e^{-U}, \quad 1/(e^{U}+1), \quad 1/(e^{U}-1) \tag{4.1}$$

If μ is $\ll E$ and if $e^{E/\mathrm{kT}} \gg 1$, the system of particles is dilute and all 3 distributions converge. The parameter μ is called the chemical potential and depends on the normalization of the system to the number of particles in the system. For dilute systems there are many open and available states for Fermi-Dirac particles. For Bose-Einstein dilute systems there is no significant clustering at low energy quantum states. Dilute systems occur at temperatures kT $\gg \mu$. Quantum statistics effects may be expected at low temperatures or with dense systems as will be explored later.

The fundamental physics resides in these weight factors. The Bose-Einstein distribution diverges as E approaches μ, and all the objects occupy the lowest quantum state. The Fermi-Dirac distribution at T = 0 has all available states occupied up to a fixed energy and the electrons are basically frozen in, without enough thermal energy to jump to an unoccupied state. For the Fermi-Dirac distribution, which has a finite μ at all temperatures, called the "Fermi Level", E_F, which will be discussed later. However, in a dilute situation, with $E \gg$ kT, all the weight factors converge, which is physically reasonable. The plots below have set the energy scale using $\mu = 1$.

```
% Energy Weighting Functions for Maxwell-Boltzmann, Fermi-Dirac and
% Bose-Einstein
% x = E/kt, normalized energies, for MB <E> = 3kT/2
```

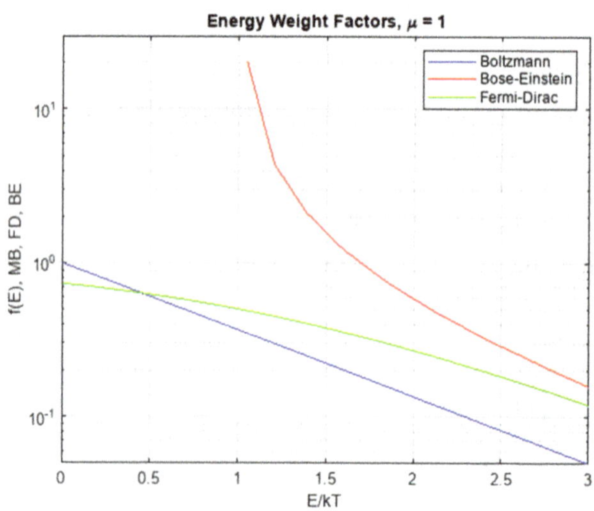

Figure 4.1: Statistical weight factors for classical particles, bosons, and fermions.

```
% The divergence at zero temperature for Bose-Einstein is very
evident.
% The convergence of all 3 for E >> kT, when the systems are dilute
is also evident
```

4.2 Phase Space

The weight functions contain the assumption about the statistics obeyed by the particles in a given application. There is a second factor which is simply the definition of the possible "phase space" occupied by the particles in the absence of these weight factors. Assuming that all momentum components are equally probable, and that there is no angular bias, the phase space in general, expressed in terms of momentum, is:

$$\mathrm{PS} = \mathrm{dp}_x\mathrm{dp}_y\mathrm{dp}_z = 2\pi p^2\mathrm{dp} \tag{4.2}$$

Phase space has the dimensions of p^3 or $(E/c)^{3/2}$ classically, or $(E/c)^3$ for photons. The spatial phase space is ignored here, assuming there is no volume containment of the system. One can easily convert to energy (kinetic) by evaluating the Jacobean of the transformation, dp/dE. For classical particles, SR particles of matter, and photons the phase space in terms of energy is, respectively:

$$\mathrm{PS} = \begin{matrix} \sqrt{2\mathrm{mE}}(\mathrm{mdE}) \\ \sqrt{(E/c)^2 - (\mathrm{mc})^2}\left(E\mathrm{dE}/\mathrm{c}^2\right) \\ E^2\mathrm{dE}/c^3 \end{matrix} \tag{4.3}$$

In quantum mechanics, there is a limit to the number of possible states in position — momentum space, which makes the normalization to the number of states contain an explicit factor $\sim(\hbar c)^3$. Quantum physics sets a limit on the number of possible states because the quantum view has "graininess".

4.3 Maxwell Boltzmann Energy, Velocity Distribution

The classical distribution of an ensemble of non-interacting particles (dilute gas) in thermal equilibrium at a temperature T is the Maxwell-Boltzmann distribution. This distribution, for kinetic energy E, or velocity v, has a specific shape and its various moments are easily evaluated using MATLAB utilities. The classical assumption is that all velocities in 3 dimensions are equally probable, $dv_x dv_y dv v_z$ which is $v^2 dv(2\pi d\Omega)$, the classical phase space for velocity. The weight displays the limiting effect of the temperature of the ensemble characterized by an exponential factor $e^{-(mv^2/2kT)}$. The translation to an energy, E, distribution is a simple evaluation of the Jacobean, dE/dv, as shown previously. The calculations are all done symbolically, using "int", "diff", "solve" and "eval". The mean and r.m.s. values in both energy and velocity are greater than the most probable values, because the distributions both have a long high energy/velocity tail that skews the distributions.

```
% moments of the Maxwell Boltzman Distribution, energy and velocity
% Initialize  - symbolic distribution - in kinetic energy E and
velocity v
dNdE = sqrt(e)*exp(-e)  % E in units of kT, not normalized here
```

dNdE = $\sqrt{e}e^{-e}$

```
dNdv = v^2*exp(-v^2/2) % v is in units of sqrt(kT/M), dNdV =
dNdE*(dE/dv)
```

dNdv = $v^2 e^{-\frac{v^2}{2}}$

```
% find norm and first 2 moments of the MB energy and velocity
distributions   % mean = 3kT/2
```

em = $\frac{3}{2}$

```
% The equipartition of energy - each dimension of a gas has a mean
energy =
% kT/2- total 3kT/2 for 3 dimensional gas.
% rms kT units
```

erms = $\frac{\sqrt{15}}{2}$

```
% most probable, <E> for 1 degree of freedom
```

emp = $\frac{1}{2}$

```
Mean % in units of sqrt(kT/M)
```

vm = $\frac{2\sqrt{2}}{\sqrt{\pi}}$

vrms = $\sqrt{3}$

```
% max of the velocity distribution
vmp = solve(ddNdv)
```

vmp = $\begin{pmatrix} 0 \\ \sqrt{2} \\ -\sqrt{2} \end{pmatrix}$

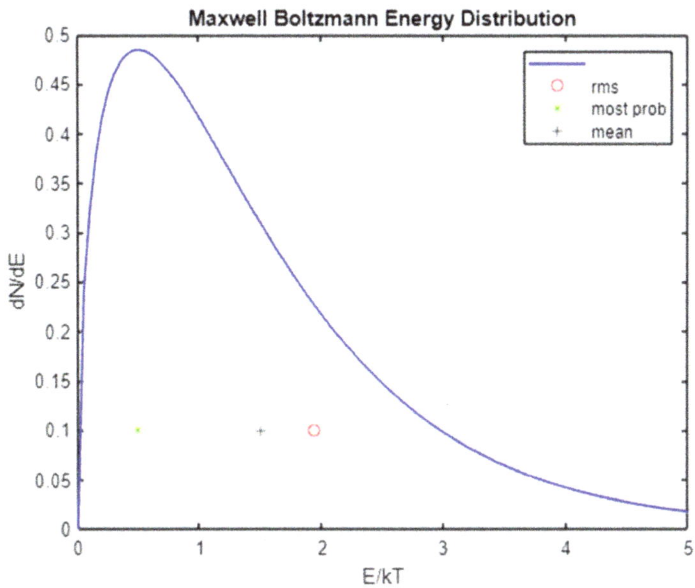

Figure 4.2: Maxwell Boltzmann classical particle energy distribution.

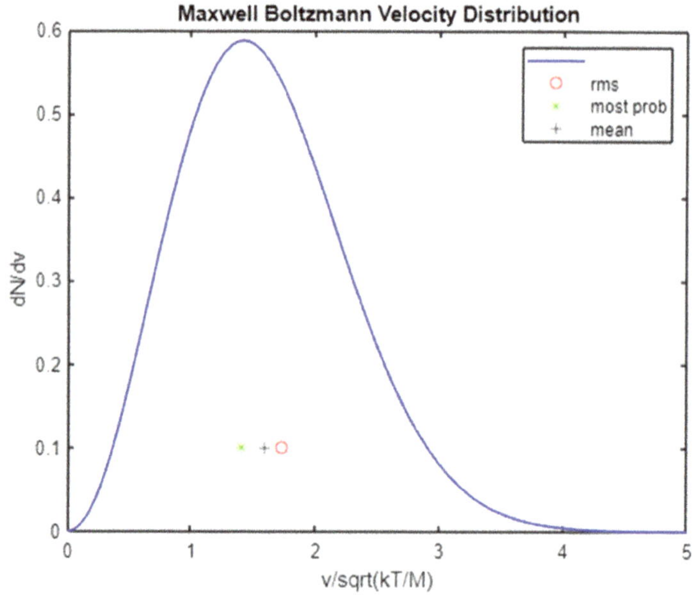

Figure 4.3: Maxwell Boltzmann classical particle velocity distribution.

4.4 Earth's Atmosphere

Assume that the ideal gas law can be applied, PV = NkT. The density of air at STP is 1.3 kg/m^3. A Pressure of 1 atmosphere is 760 Torr = 101 kPa. The Maxwell-Boltzmann results are $\langle E \rangle$ = 3kT/2, $\langle v^2 \rangle$ = 3kT/m. The value of kT at STP is ~1/40 eV. At that temperature a gas of protons has a thermal velocity of β of 9×10^{-5}, $v \sim 2.7$ km/s.

It is notable that there is very little Helium on Earth. That seems to contradict the model of element formation in the Sun, where a first generation star would be 78% hydrogen and 22% Helium, with almost no heavier elements, as will be covered later. The resolution of the issue is that Helium has a velocity distribution rather faster than the O2 and N2 which make up most of the atmosphere. Helium on the other hand can thermally fluctuate to have a velocity larger than the escape velocity. In addition Helium is a "noble" gas, and is not captured in more complex molecules as are Oxygen and Nitrogen. The plot below shows the differences. Note the many decades of vertical log scale.

```
% Ideal Gas Law and Earth's atmosphere
% now a few MB distributions of energy and velocity - O2 example so
m = 32*mp
kT(1) = 1 ./40;       % approx for 300 deg K - eV
Ao = 32;               % O2 mol weight
Ahe = 4;
mp = 0.938 .*10 .^9 ; % p/n mass in eV
c = 3.0 .*10 .^5    ;  % c in km/sec
g = 9.8; % accel at earth surface m/sec^2
re = 6.378 .*10 .^6; % earth radius - m
ve =sqrt(2.0 .*g .*re); % escape velocity ~ 11.2 km/sec
ve = ve ./1000 ;
velHe = sqrt(2.0 .*kT(1) ./(Ahe .* mp)) .* c    % He velocity in
km/sec
```

```
velHe = 1.0952
```

```
velO2 = sqrt(2.0 .*kT(1) ./(Ao .* mp)) .* c    %  O2 velocity in
km/sec
```

```
velO2 = 0.3872
```

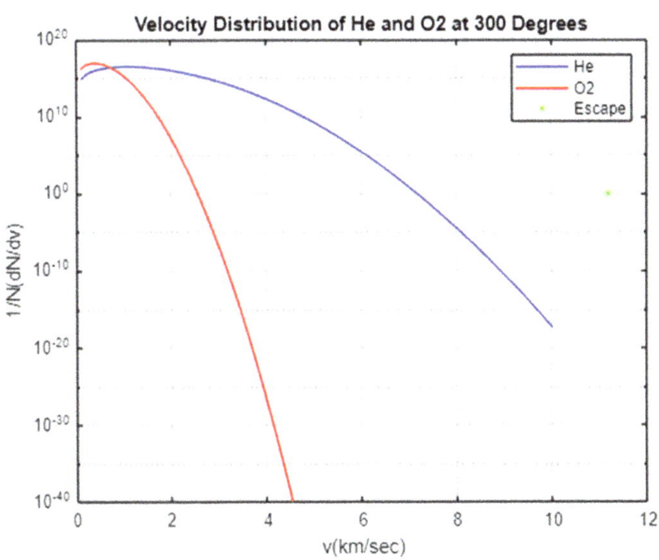

Figure 4.4: Maxwell Boltzmann classical particle velocity distribution for atmospheric Helium and Oxygen. Note the many decades on the vertical scale.

4.5 Planck Blackbody Distribution

Before the distinction in statistics between Fermi-Dirac and Bose-Einstein statistics of half integer and integer spin particles, classical thermodynamics was altered by Planck and a different energy distribution was postulated to match the blackbody radiation experimental data. An ad hoc constant, the Planck constant, $\hbar$, was defined to connect the temperature, photon frequency, and the energy E, $E = \hbar\omega$, and the weight factor was altered. In this case there is no normalization which is specified since the number of photons is not constrained and the chemical potential, μ, which sets the normalization, is 0. The specific definition for the chemical potentials and the distributions are deferred. The Planck distribution is simply the Bose-Einstein distribution with zero chemical potential. The photon energy is E and the black body system energy U is distributed in E. The classical Boltzmann distribution for gases and liquids was discussed previously.

```
% moments of the Planck distribution, energy density for a photon
gas
% Initialize  - symbolic distribution - in energy E
syms dUdE k T e IdUdE IEdUdE IE2dUdE EM E2M em e2m erms ddUdE emp
norm dUdEMB
dUdE = e^3/(exp(e)-1) % E in units of kT
```

dUdE = $\dfrac{e^3}{\mathrm{e}^e - 1}$

```
% find norm and first 2 moments of the Planck energy a
% distributions
IdUdE = int(dUdE,0,inf)
```

IdUdE = $\dfrac{\pi^4}{15}$

```
norm = dUdE/IdUdE;
IEdUdE = int(e*dUdE,0,inf);
IE2dUdE = int(e*e*dUdE,0,inf);
%
em = IEdUdE/IdUdE        % mean
```

em = $\dfrac{360\,\zeta(5)}{\pi^4}$

```
e2m = IE2dUdE/IdUdE;    % mean squared
EM = em*k*T;
E2M = e2m*k*k*T*T;
erms = sqrt(e2m)        % rms
```

erms = $\frac{\pi\sqrt{21}\sqrt{40}}{21}$

```
ddUdE = diff(dUdE);
emp = solve(ddUdE)      % most probable, special function W
```

emp = $W_0(-3e^{-3}) + 3$

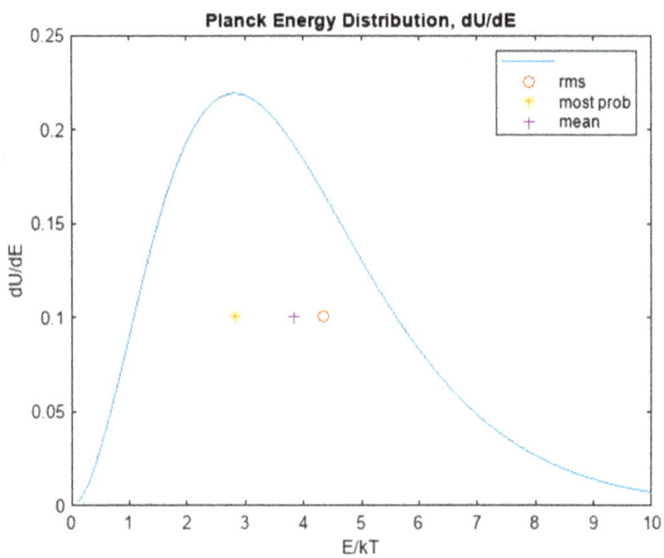

Figure 4.5: Planck bosonic particle energy distribution, without chemical potential.

4.6 Bose-Einstein and Fermi-Dirac Chemical Potential

The reader will have noticed the use of E for kinetic energy in this section rather the T. That is, alas, conventional, as is the use of cgs units. The Maxwell-Boltzmann distribution for kinetic energy can be integrated to relate the total number N to the temperature and the as yet undetermined normalization factor A. This normalization is purely classical physics.

$$\mathrm{dN/dE} = A\sqrt{E}e^{-E/\mathrm{kT}} \cdot N = A\sqrt{\pi}(\mathrm{kT})^{3/2}/2. \quad (4.4)$$

The density of possible quantum states, $n = N/V$ needs to be used in the following cases. There is typically 1 quantum state possible per energy of $\mathrm{dn/dE} \sim 1/\hbar^3$. More correctly the number density scales as $\sqrt{E}$ as it does classically, but now the proportionality is set by the quantum "graininess" to be $\mathrm{dn/dE} = (\sqrt{2}m^{3/2}\sqrt{E})/\pi^2\hbar^3$. In classical physics there was a kinetic and potential energy, T and V. For quantum statistics, there is conventionally E and the "chemical" potential μ which appears in the weight factors as $E - \mu$, hence the name.

For the FD and BE distributions the integrals over the distribution functions are considerably more difficult and embody the very different physical properties of fermions and bosons. The normalized distributions define an energy parameter. μ, called the chemical potential. That potential is defined so that the FD and BE distributions with E replaced by $E - \mu$ are properly normalized. The chemical potential depends on temperature, T.

In the Bose-Einstein case the integrals involve the Riemann zeta functions, which are available numerically in MATLAB as "zeta(z)". Ignoring constants for now the basic integral for the number density $n = N/V$ results in the Riemann zeta function of index 3/2, while for the energy the index is 5/2. Choosing a substance, the number density, n, and number of spin states, s, are fixed, typically 2 for elementary spin 1/2 fermions and 3 for spin 1 vector bosons. A explicit solution for $\mu(T)$ is found by searching at fixed T until a $\mu(T)$ is found which satisfies the normalizing expression for n. The energy density, $u = U/N$, at that value of T can then be explicitly calculated, knowing μ. The exponential of μ/kT is proportional to both number and energy, and thus sets the normalization.

$$
\begin{aligned}
&\mathrm{dn} \sim \sqrt{E}/[e^{(E-\mu)/\mathrm{kT}} - 1]\mathrm{dE} \\
&n = s(\mathrm{mkT}/2\pi\hbar^2)^{3/2}\zeta_{3/2}(e^{\mu/\mathrm{kT}}) \\
&u = (3/2)\mathrm{kT}(\mathrm{mkT}/2\pi\hbar^2)^{3/2})\zeta_{5/2}(e^{\mu/\mathrm{kT}})
\end{aligned} \tag{4.5}
$$

The Planck distribution for blackbody radiation is just the Bose-Einstein distribution for vector photons with a vanishing chemical potential. The number of blackbody photons is unconstrained unlike the number of atoms in a box, which implies that $\mu = 0$. The expression for u approaches the classical Boltzmann result at high temperatures. The parameter s is the number of spin states of the element of mass m. Also plotted below is an estimate of the critical energy. That energy is defined to occur when the deBroglie wavelength, $\lambda_{\mathrm{dB}} = 2\pi\hbar/p$, for the specific case of Helium is comparable to the inter atomic spacing, because quantum effects such as superfluid He behavior, are expected to occur at approximately these temperatures. The critical temperature is estimated to be:

$$T_c = (2\pi\hbar^2/\mathrm{mk})[n/s\zeta_{3/2}]^{2/3} \tag{4.6}$$

The plot below for He is the chemical potential in kT units and the estimate of the critical temperature as the red dot. In fact, Helium becomes a superfluid at 5.8 degrees K, which is close to the estimate of 2.8 degrees. The chemical potential for Helium is small for low temperatures and then rises with T above about 10 degrees.

```
% Chemical Potential for BE and FD
% For BE, He4 Nuclei While for FD, e in Li
% constants
rho = 0.125 ;  % He density gm/cm^3
rho = rho .*10 .^3;      % density in kg/m^3, liquid
No = 6.02 .*10 .^26;     % Avagadros # - MKS units
AHe = 4;
N_V = No .*rho ./AHe;    % number density
h = 6.6 .*10 .^-34;      % Planck const j-sec
hbar = h ./(2.0 .*pi);
k = 1.38 .*10 .^-23;     % Boltz const - j/K
m = 1.67 .*10 .^-27;     % p mass kg
zet = zeta(1.5); % Riemann zeta 3/2
Ec = y .*((N_V ./zet) .^0.6666) ;    % critical energy, spin zero
Ecev = Ec ./(1.6 .*10 .^-19); % in eV - convert joule
% Critical Energy and Temperature in He: Ecev, Tc)
% find the chemical potential given the number density - explicitly
```

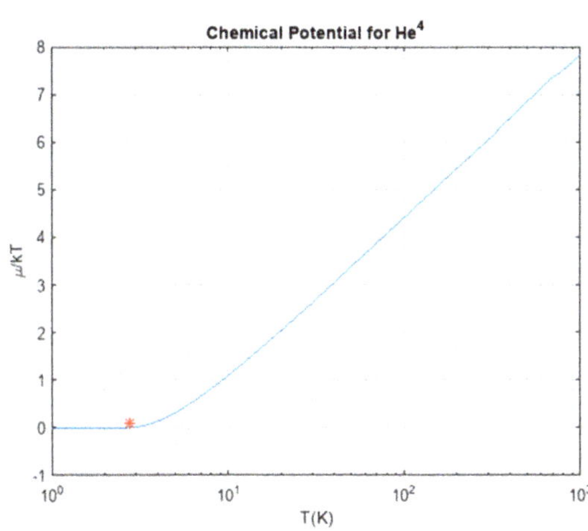

Figure 4.6: Chemical potential for bosonic Helium as a function of temperature. The estimated critical temperature is the red dot.

In the Fermi-Dirac case the function is simplest at $T = 0$ but complex at finite T. Most states at low temperature are "frozen" by the Exclusion Principle with no vacant adjacent states since all states with energy less than the Fermi level are filled at $T = 0$. Only those within energies ~kT of the Fermi level can gain energy ~kT because of the existence of vacant states nearby. In the lowest level of approximation, the chemical potential is approximately the Fermi level at $T = 0$. The momentum, or wave vector depends on the 1/3 power of the number density.

$$
\begin{aligned}
n &= N/V = \int_0^{E_F} \sqrt{E}\mathrm{dE} \sim E_F^{3/2} \\
\mu &= E_F(0) = \hbar^2/2m_e(3\pi^2 n)^{2/3} = \hbar^2 k^2/(2m_e) \\
k &= (3\pi^2 n)^{1/3}
\end{aligned}
\tag{4.7}
$$

In general the exact expression for the Fermi chemical potential at finite temperature requires the evaluation of the integral $n \sim \int_0^\infty \sqrt{E}/[1 + e^{(E-\mu)/\mathrm{kT}}]$. In what follows an approximate expansion is made as a way to estimate the Fermi Level as a function of T. An exact numerical computation will be done in the following separate script. The example below is for electrons in Lithium. The chemical potential is equal to the Fermi level at low temperatures, indicating that the electrons are "frozen" at low temperatures. The red dot is the Fermi level at $T = 0$ in temperature units. The different dependence of the chemical potential on temperature for bosons and fermions is a reflection of the different statistics obeyed by the 2 types of matter.

```
% now Lithium for e in metals - FD
rho = 0.534 ;  % Li density gm/cm^3
rho = rho .*10 .^3;      % density on kg/m^3
No = 6.02 .*10 .^26;     % Avagadros # - kg
ALi = 6.94;              % atommic weight
N_V = No .*rho ./ALi;    % number density
me = 9.1 .*10 .^-31;     % me in kG
c1 = 1.6 .*10 .^-19 ;    % eV to j, electron charge
% compute T = 0 Fermi Level
kf = (3.0 .*pi .*pi .*N_V) .^0.33333;
ef = (hbar .*kf) .^2 ./(2.0 .*me);
ef = ef ./c1; % joule to eV
tf = ef .*40 .*300; % assocaited temperature
% Fermi Level and Temperature e in Li: ef(eV), tf(oK)
```

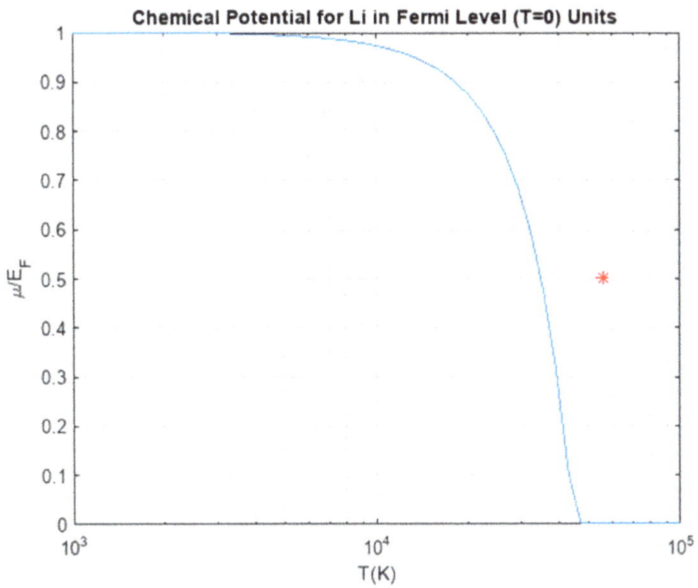

Figure 4.7: Chemical potential for fermionic Lithium as a function of temperature. The $T = 0$ Fermi energy level is the red dot.

4.7 Fermi Chemical Potential

In order to evaluate the Fermi-Dirac chemical potential at finite temperature, numerical integrals are needed and the MATLAB utility for numerical integration, "quad", is used. The chemical potential μ is defined to be the finite temperature Fermi level. Ignoring constants for now the number density as a function of energy defines the chemical potential μ. while at $T = 0$, $\mu = E_F = (\hbar k_F)^2/2m$, $k_F = (3\pi^2 n)^{1/3}$. The $T = 0$ wave vector k depends only on the number density. As in the Bose-Einstein case, integrating over all energy E to find the number density, n, and equating that to a term depending on μ and T enables a computation of $\mu(T)$.

$$\mathrm{dn/dE} = \sqrt{E}/[\mathrm{e}^{(E-\mu)/\mathrm{kT}} + 1] \tag{4.8}$$

As an aside, it is a question why the world appears solid to us. After all it is known that atoms have nuclei with most of the mass of the atom and with small sizes, ~ fermi or 10^{-15} m. Atoms are of size 10^{-10} m. Atoms have electrons in orbits and the electrons have Compton wavelengths ~50 times smaller than the orbital sizes. That means the "solids" are mostly empty space. The Fermi Exclusion Principle and Fermi pressure imply that matter has a bulk modulus $B = 5p_F/3$, $1/B = -(1/V)(\mathrm{dV}/\mathrm{dp})$ the fractional change in volume with pressure. An estimate for B using the $T = 0$ Fermi level which only depends on the number density, is then $10^{10}\mathrm{nt/m}^2$ for $Z = 10$. That is, indeed, the right order of magnitude. It is the Fermi pressure that gives us the illusion of solidity.

```
% dn/dE ~ sqrt(E)/(exp((E-EF)/kT) + 1), EF Depends on Density -
Chemical Potential
% At T = 0, E_F = (hbar*k_F)^2/2m, k_F=(3*pi*pi*N/V)^0.333
hbarc = 2000.0;              % hbarc in eV*A
mec2 = 511000.0;             % electron mass in eV
k = 1.0 ./(40 .*300);        % Boltzmann const, kT - 1/40 eV at room
temp = 300 degrees absolute
% units of Angstroms or number density in 10^23/cm^3
% first find the T = 0 limits where the integrals are closed form
nz = ((EF .^1.5) .*C .*2) ./3; % zero temp EF
```

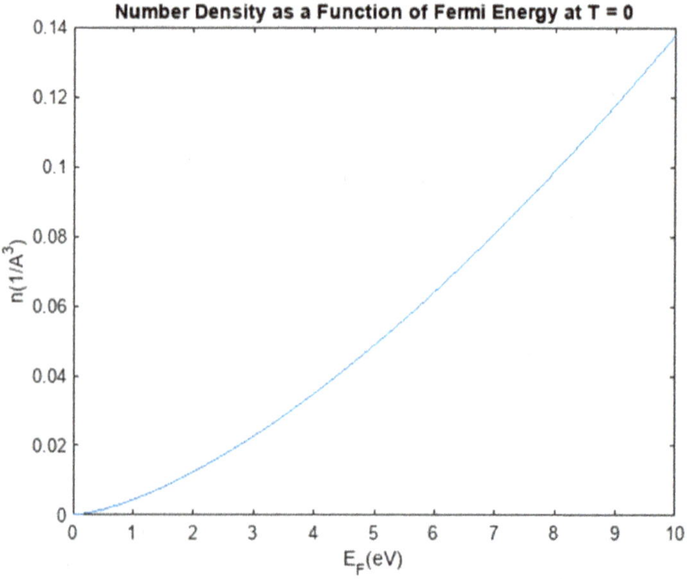

Figure 4.8: Number density, n, as a function of Fermi energy level at $T = 0$.

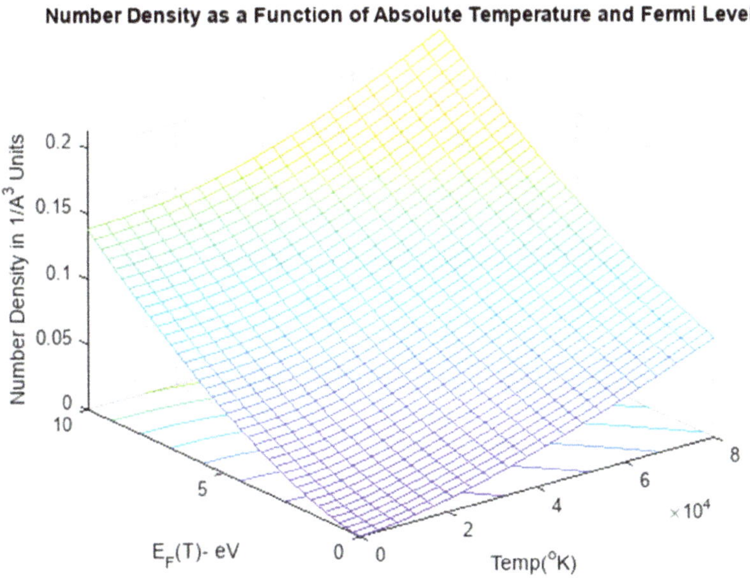

Figure 4.9: Number density, n, as a function of Fermi energy level and temperature, T.

The energy level rises with temperature and number density. More empty states are available at elevated temperatures. The zero temperature result scales as the 1/3 power of the number density. For room temperatures and below the changes in Fermi energy from the zero temperature result is normally rather small, so the approximation of using the $T = 0$ Fermi level is often used.

4.8 Pressure and Molecular Collisions

The Maxwell-Boltzmann distribution applies to a dilute gas of non-interacting particles. It can be extended by looking at gas collisions among themselves and with the walls containing the gas. The latter collisions are the microscopic explanation for the pressure, and imply that pressure is proportional to temperature, as expressed by the ideal gas law. That dependence can be seen by varying the temperature of the 2 dimensional "gas". The gas is distributed by Monte Carlo methods to be uniform in the box and to have a 2-d Maxwell-Boltzmann energy distribution. The reported pressure is $\sim$ found to be proportional to temperature. The user chooses the temperature. Even with low statistics in the number of molecules, the momentum transferred to the walls scales ~ linearly with kT for kT < 5. At higher temperatures, the effects of the collisions between the particles changes the linear behavior. The particle collisions among themselves and not the walls are the cause of the viscosity, which scales as $\sqrt{\mathrm{kT}}$. as will be discussed momentarily. This code is simply a "hard sphere" collision model, where molecules scatter isotropically if they overlap in their radial sizes. The user picks the temperature and can view the time evolution of the gas.

```
% Model with gas-gas collisions -> viscosity
% first setup the "gas" - points with a 2-d MB energy distribution
nmol = 80; % number of gas molecules
aa = 0.025; % keep the gas dilute, % size of molecule = aa = 0.05
%  total 2-d  N possible in 1 x 1 area = 400, still ~ dilute
ntime = 50; % Number of Time Steps
% pick a "temperature"- mass of the gas points defined to be 1
```

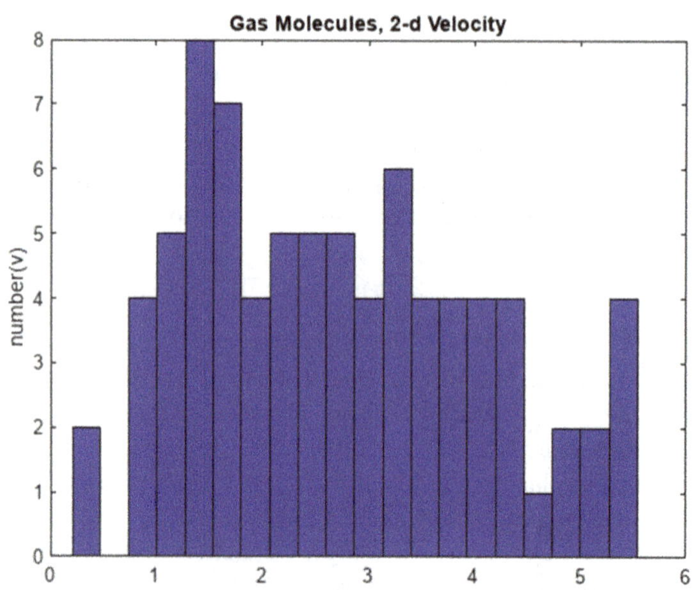

Figure 4.10: Histogram of the 2-d classical velocity distribution for the simple gas model.

```
pm = mean(p)
```

```
pm = 2.7393
```

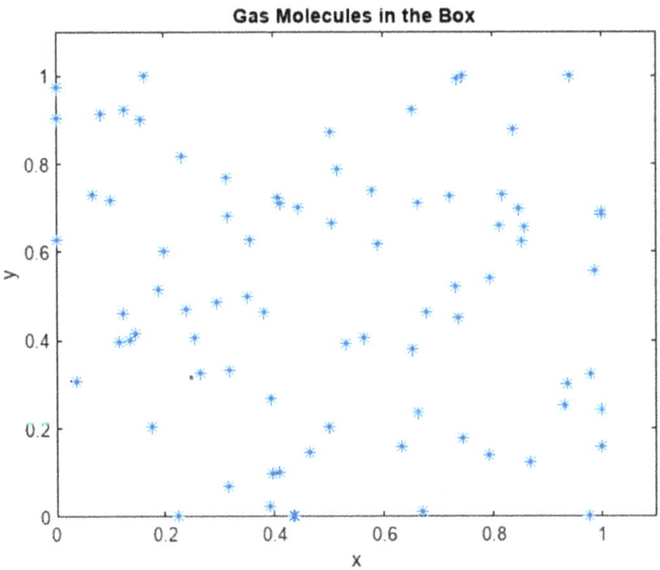

Figure 4.11: Snapshot of the 2-d position distribution for the simple gas model used in tracking wall and molecular collisions.

```
% play back the movie as desired
ncoll % Number of Wall Collisions
```

```
ncoll = 633
```

```
pcoll % Momentum Impulse to Walls
```

```
pcoll = 3.3341e+03
```

4.9 Gas Viscosity

Gas viscosity was mentioned previously in the context of "drag" on a rocket, and η was the symbol used for viscosity. Viscosity arises from the collision between molecules and is absent in the Maxwell-Boltzmann formulation which assumes a non-interacting. dilute gas. The dimension for viscosity are Pa $* s$. The scale for viscosity, for air at STP or 101 kPa, is $\eta \sim 18(\mu\text{Pa} * s)$. In comparison for water the viscosity is $10^{-3}\text{Pa} * s$, about 100 times larger, since the medium is denser and has more molecular collisions per second. The molecular collisions are modelled in the hard sphere approximation. In that model a molecule of radius R collides with another with a cross section equivalent to a point molecule colliding with a fictitious sphere with radius 4R. The cross section for collisions is $\sigma = 16\pi R^2$ and the mean free path between collisions is $\langle L\rangle \sim 1/(n\sigma)$ where the number density is n. The estimated mean time between collisions is $\tau \sim \langle L\rangle/\langle v\rangle$. The result for viscosity scales as $\sqrt{T}$, increasing with temperature and decreasing with the cross section.

$$\begin{aligned} &\langle \nu\rangle = \sqrt{8\text{kT}/\pi m}, \quad n\sigma\langle v\rangle = 1/\tau \\ &\langle L\rangle \sim 1/n\sigma, \quad \eta = (1/3\sqrt{2})m\langle v\rangle/\sigma = (2)/(3\sqrt{\pi})\sqrt{\text{mkT}}/\sigma) \end{aligned} \tag{4.9}$$

The code below displays a variety of numerical results. They should give the user a feeling for collisions in gases and liquids. The user can choose the atmospheric element and the gas or liquid phase. The data for the viscosity of H2 gas at STP is $9.5 \times 10^{-5}\,\text{gm}/(\text{cm} * s)$. The estimated value which is displayed is within a factor of 2.

```
% Transport and viscosity and the Maxwell - Boltzmann distribution
% some data for molecular gases/liquids
A = [2 4 28 32];    % molecular, diatomic
rhog = [0.084 0.166 1.165 1.332]; % gas density in gm/1000 cm^3
rhol = [0.071 0.125 0.807 1.141]; % liquid density in gm/cm^3
ra = [4.6 1.2 0.7 0.56] ;          % atomic/molecular radius in A
%
iA  = 1 ; % Element,H,He,N,O
iGas = 1; % State, gas,liquid
% effective cross section is pi*diameter^2
sig = pi .*4 .* (rAA .* 10 .^-16);     % rAA is in A, go to cm
% find mean thermal velocity
```

```
mp = 938.0 .*10 .^6;      % p mass in eV
c = 3.0 .*10 .^10;       % c in cm/sec
No = 6.02 .*10 .^23;     % Avagadros # - gm and cm
ktm = (1.0 ./(40.0 .*mp .*AA));
v = sqrt((8.0 .*ktm) ./pi);
v = v .*c;               % mean thermal velocity in cm/sec
n = (No .*rho)./AA;      % number density of atoms/cm^3
sig % Cross section for gas scattering = pi*(2*r)^2 = (cm^2)
```

```
sig = 5.7805e-15
```

```
v   % Mean thermal velocity at 300 oK = sqrt(8*kt/pi*mp*A) ,(cm/sec)
```

```
v = 1.7476e+05
```

```
n   % number density = Number of molecules/cm^3 = No*rho/A
```

```
n = 2.5284e+19
```

```
nc = (n .*n .*sig .*v) ./sqrt(2) % collisions/cm^3*sec - total
```

```
nc = 4.5666e+29
```

```
Nc = nc ./n   % collisions of a single molecule - collisions/sec
```

```
Nc = 1.8061e+10
```

```
tau = 1.0 ./Nc  % mean time between collisions
```

```
tau = 5.5368e-11
```

```
lmean = tau .*v   % mean free path (cm) between collisions
```

```
lmean = 9.6761e-06
```

```
eta % Viscosity = (rho*lmean *v)/3 =  (gm/(cm*sec))
```

```
eta = 4.7348e-05
```

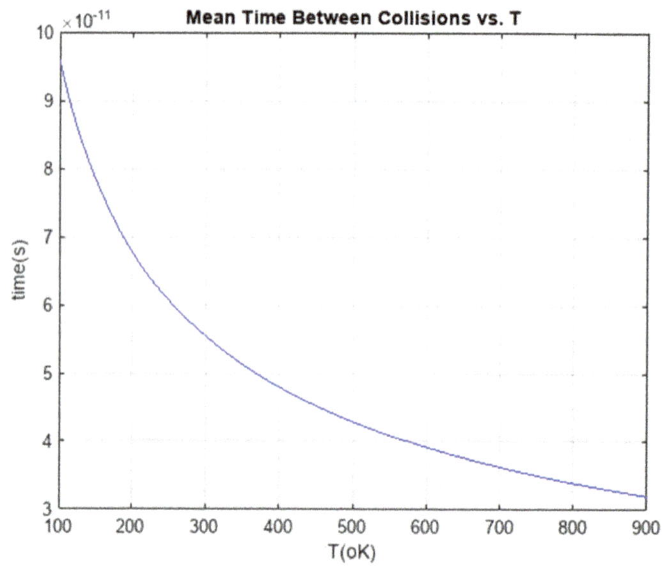

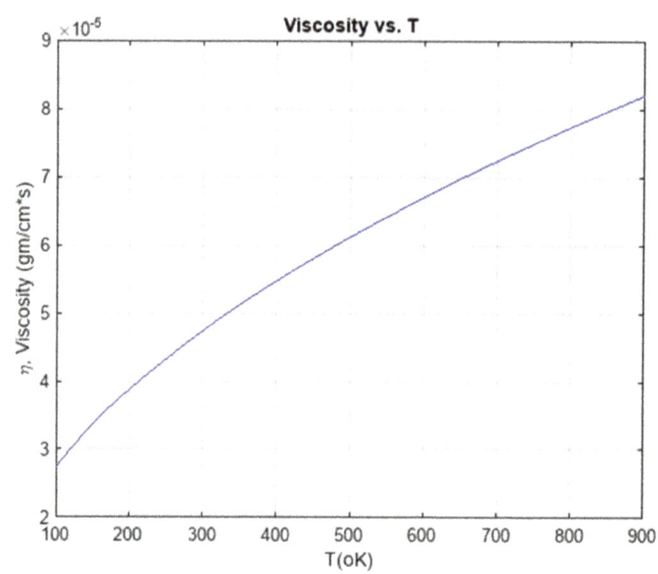

Figure 4.12: Top: Mean time between collisions as a function of temperature. Bottom: Estimated viscosity as a function of temperature.

4.10 Stoke's Law and Viscosity

Consider the case of dropping an object through a gas or fluid which has viscosity. In this case the fluid is a liquid with a known viscosity, η, with an acceleration which is therefore velocity dependent. Assume the gravitational force is constant with acceleration g reduced by the buoyancy of the object to a value a. The 1 dimensional vertical equation of motion is then shown below. Before computing, one can notice that the acceleration is 0 when a velocity, the constant terminal velocity, $v_t = a/b$, is reached. Knowing the density of the fluid and the terminal velocity of the object means one can experimentally measure the viscosity.

$$\frac{\mathrm{d}^2}{\mathrm{d}t^2}y = a + b\frac{\mathrm{d}}{\mathrm{d}t}y = 0 \tag{4.10}$$

Specifically look at a sphere released at rest into water. The larger the viscosity the slower the terminal velocity. For the specific case of a sphere of density ρ_s falling through water with density ρ_w, the terms are $a = (\rho_s - \rho_w)g/\rho_s$ correcting for buoyancy, and $b = -(9\eta)/(2\rho_s R_s^2)$ for a sphere of radius R_s in the liquid with viscosity η. The steam lines for the flow through the water are determined by the flow potential ψ with stream lines tangent to the local velocity at a location specified by the y and r coordinates. The potential ψ falls off with r far from the moving sphere and initially grows with y as the sphere picks up speed. The code uses the MATLAB utility "streamline".

$$\psi(y, r) = -y[1 + (R_s/r)^3/2] \tag{4.11}$$

The terminal velocity measurement determines the viscosity. As mentioned previously, the viscosity depends on the molecular size, R since the scattering, which goes as the radius squared, defines the viscosity. Therefore, a simple macroscopic measurement of a terminal velocity determines the size of the molecules whose collisions are the root cause of the viscosity, $\eta \sim 1/\sigma$, $\sigma \sim R^2$.

```
ystoke = dsolve(D2y == a+b*Dy, y(0) == 0, Dy(0) == 0)
```

ystoke = $-\dfrac{a - ae^{bt} + abt}{b^2}$

```
vstoke = diff(ystoke)
```

vstoke = $-\dfrac{ab - abe^{b_t}}{b^2}$

```
R = 1e-2;          % radius of sphere, m
rho = 7.9e3;       % density of sphere kg/m^3 - Fe
rhoF = 1e3 ;       % density of fluid (water)
drho = rho-rhoF;
mu = 5.0e-3; %viscosity of fluid = water, 1e-3 Pa*sec
g = 9.8; % gravity
a = drho ./(rho .*g) ;
b = -(9 .*mu) ./(2.0 .*rho .*R .*R);
vterm = a ./b ;      % terminal velocity
```

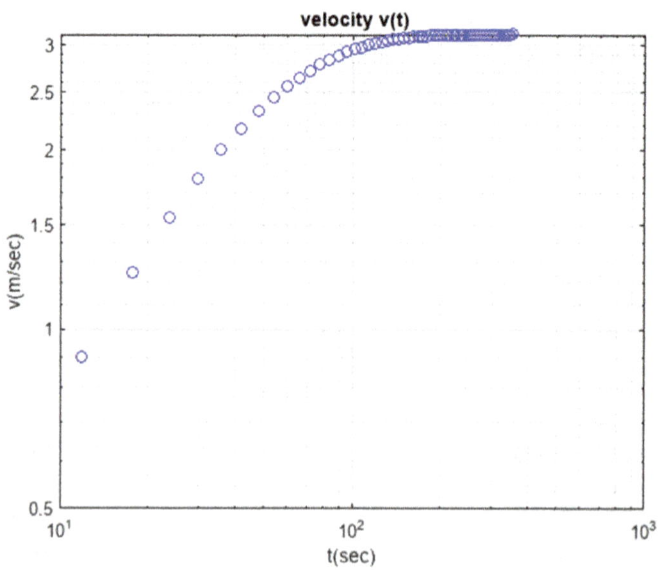

Figure 4.13: Velocity as a function of time for a sphere falling in a viscous fluid. The limiting velocity is very clear as is the falling behavior of the acceleration under gravity.

```
% streamlines at terminal velocity - in frame at rest on sphere
% u = vx, v = vy in 2-d for sphere in uniform flow v = 1,
% flow potential
phi = -y.*(1.0 + (R ./r) .^3 ./2.0);
```

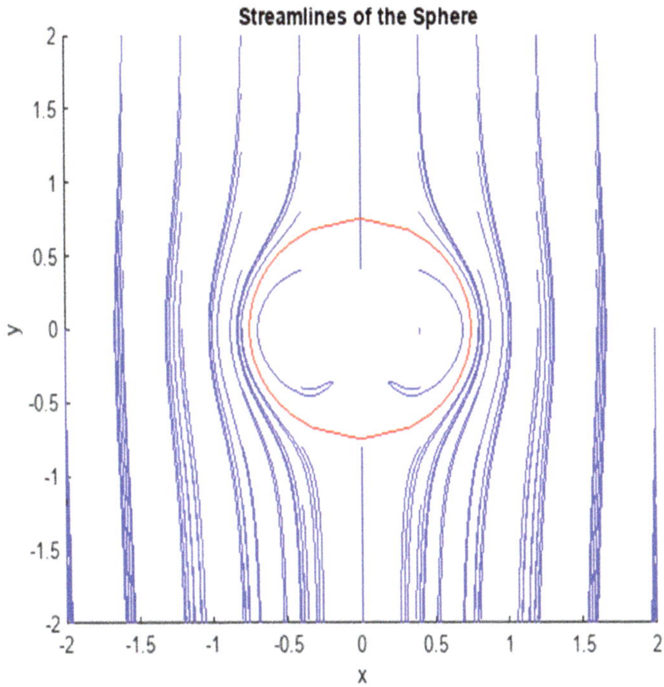

Figure 4.14: Velocity streamlines for a falling sphere at terminal velocity.

4.11 Viscous Flow

The simplest flow has no interactions within the medium. The flow is laminar and the velocity of the medium vanishes on the constraining boundaries of the fluid. Internal friction can, however, be added if there is no turbulence by specifying the viscosity of the liquid. The velocity, v, of the fluid, with viscosity η in a circular pipe of radius R and length L driven by a pressure P is:

$$\begin{aligned} \mathrm{dv/dr} &= -2\,\mathrm{Pr}\,/\eta L \\ v(r) &= (P/L)(R^2 - r^2)/\eta \end{aligned} \tag{4.12}$$

The velocity of the liquid flow depends on the pressure per unit length and inversely as the viscosity. It is largest at the center of the cylindrical pipe and vanishes at the pipe walls. The overall volume flow is: $\mathrm{dV/dt} = \pi R^4(P/L)/(2\eta)$ with a factor R^2 from the velocity and a factor R^2 due to the cross section of the pipe. In the code below the pressure is set to 1 Atm and the length of the pipe is 1 km, with a pipe radius of 5 cm. The user picks the factor by which the viscosity of water is scaled. The liquid velocity and volume flow rate both scale inversely with viscosity and linearly with the pressure per unit length.

```
% laminar flow in a cylindrical pipe
% Frictional Force in Fluid, viscosity Balances Pressure Force
% dv/dr = Velocity Gradient = -2Pr/(eta*L)
P =  1; % Pressure in Atm
P = P .*10 .^5;   % 1 Atm ~ 10^5 Pa
eta = 1006 .*10.^-6; % uPa*sec  water viscosity
fac = 1 % Factor for liquid viscosity w.r.t. water
```

```
fac = 1
```

```
eta = eta .*fac
```

```
eta = 0.0010
```

```
dVdt = pi .*P .*(R .^4) ./(2 .*eta .*L)  % volume flow through pipe.
```

```
dVdt = 0.9759
```

```
% Volume flow (m^3/sec) dVdt
```

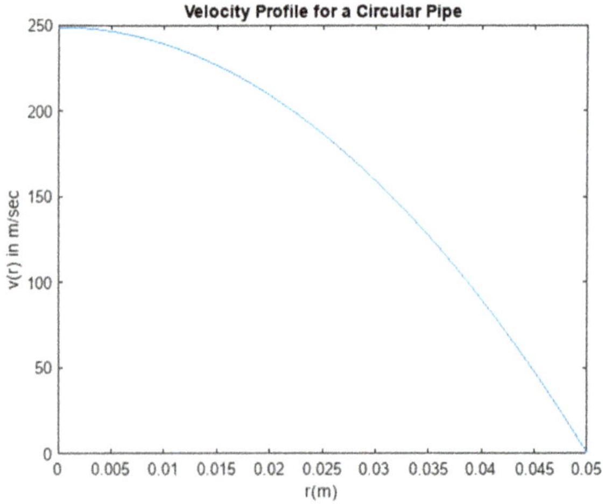

Figure 4.15: Velocity as a function of radius for viscous flow in a cylindrical pipe.

4.12 Fluid Flow, Streamlines

For incompressible fluid flow, $\vec{\nabla} \bullet \vec{v} = 0$. The math is the same Laplace equation as the electric field without any charge density. Away from the sources; the divergence of the field is zero. That means that the vector field can be derived from a scaler potential. In the case of fluid flow using the velocity potential ψ and a 2 dimensional velocity $\vec{v} = (u, v)$ with streamlines defined to be parallel to the velocity vector.

$$\vec{v} = \nabla\psi \tag{4.13}$$

Solutions for a few different sources are known, just as in the case of electrostatics, for example a point source. Look at the sources of flow and sinks, analogues of positive and negative point charges. The choices for the user to consider are for a constant velocity, a point source or a sink. The stream lines are analogous to the electric fields, and the "gradient" MATLAB utility can again be used, or alternatively the utility "streamline" can also be used.

```
% script on velocity potential, vx, vy, and streamlines
% grid for evaluation, 4 x 4 step 0.2
% constant velocity flow along x, vx = u, vy = v by convention
uc = 1;
vc = 0;
% velocity from grad of flow potential phi
phic = uc .*x + vc .*y; % vx = uc, vy = vc
% now add a point source at (0,0)
Q = 5.0; % source strength
% source/sink, familiar from electric field and potential for a point charge
vrso = Q ./(2 .*pi .*r); vthso = 0; phiso = (Q .*log(r)) ./(2 .*pi);
vrsi = Q ./(2 .*pi .*r); vthsi = 0; phisi = -(Q .*log(r)) ./(2 .*pi);
```

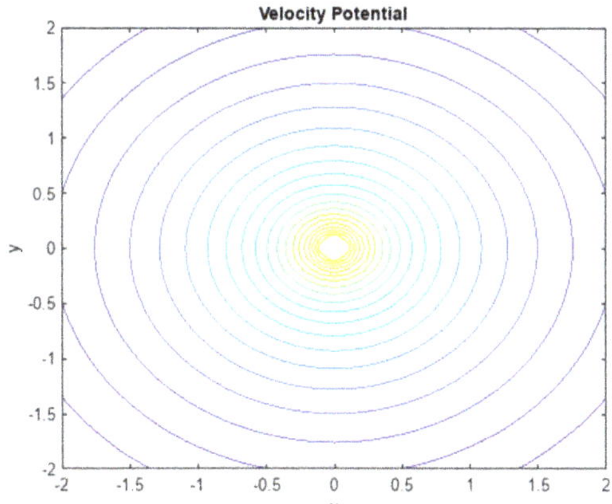

Figure 4.16: Velocity potential contour for a point source of flow.

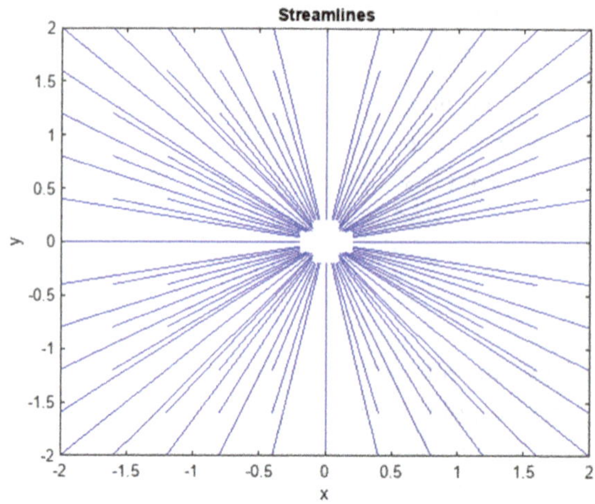

Figure 4.17: Velocity streamlines for a point source of flow.

4.13 Heat Equation

The equation governing heat flow has both space and time coordinates which is a new situation to be explored. The numerical solution is also made on a grid, as was done for static problems in electromagnetism. Now, however there is an issue for the stability of the solutions. Consider the simplest case of heat flow in 1 spatial dimension, x, defined by the temperature, T, flow in time t.

$$\frac{\mathrm{d}^2}{\mathrm{d}x^2}T - \frac{\mathrm{d}}{\mathrm{d}t}T \tag{4.14}$$

The grid approximation in space is $\frac{\mathrm{d}^2}{\mathrm{d}x^2}T \sim [2T_i - (T_{i+1} + T_{i-1})]/\mathrm{d}x^2$ as discussed previously. The variation in time has several choices. The simplest is called forward, $\frac{\mathrm{d}}{\mathrm{d}t}T \sim T_{i,n+1} - T_{i,n}$ where n refers to a grid location in time. Now there is a question of the stability of the solution. In 1 dimension, in the forward time step case: $T_{i,n+1} \sim T_{i,n}(1 - 2F_o) + F_o(T_{i+1,n} + T_{i-1,n})$, $F_o = \mathrm{d}t/\mathrm{d}x^2$. The grid solution mixes space and time variations characterized by the parameter F_o and clearly that parameter needs to be less than $1/2$ for numerical stability. A larger number of spatial dimensions increases the requirement on F_o to maintain stability. For 2 spatial dimensions the magnitude of F_o is halved.

The stability issues are explored very simply in a look at the stability of 1 dimensional temperature flow. There is a fixed spatial grid, and the user can vary the granularity of the temporal grid. The solutions will, as expected, begin to lack stability when Fo is ~0.5. As with other nonlinearities, the onset is very sharp for small variation of the parameter Fo. These considerations will also apply for wave equations and quantum mechanical equations for time development of a system.

```
% Solution for 1-d spatial, 1-d time Heat Eq. Study of stability,
temp T
% Finite Difference for d2/dx2:  2Ti = (Ti+1 + Ti-1) /dx^2 grid in
1d, generalize for
% more spatial dimensions
% for d/dt forward: d/dt = (Ti,n+1 -Ti,n)/dt for iteration n+1
          % backward:     = (Ti,n - Ti,n-1)/dt
          % central       = (Ti,n+1-Ti,n-1)/dt
% forward and backward errors ~ O(dt), central error ~ O(dt^2)
% for forward dt in 1 spatial d: Ti,n+1 = Ti,n(1-2Fo)+Fo(Ti+1,n
+Ti-1,n)
% Fo = dt/dx^2. For 1 space d stability if Fo < 0.5. For 2-d Fo =
2dt/dx^2
dt =  0.000196;
Fo = dt ./del .^2
```

```
Fo = 0.5098
```

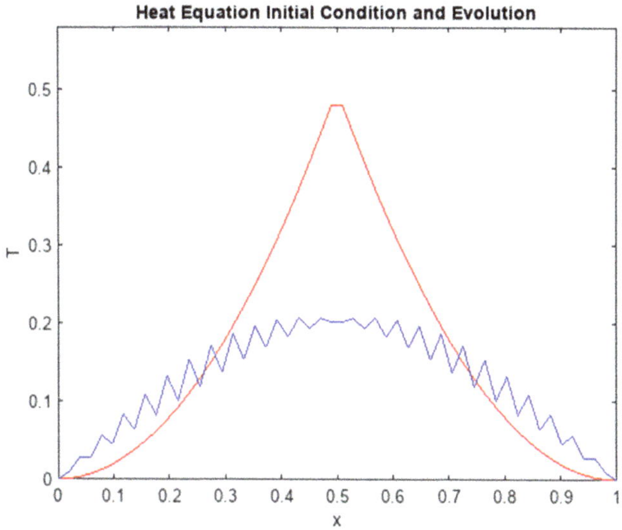

Figure 4.18: Temperature profile as a function of position for 1-d heat flow with initial temperature profile at $t = 0$ in red.

```
% The movie can be played back as desired. The onset of instability
at Fo =
% 0.5 is quite rapid.
```

4.14 Heat Diffusion

MATLAB has a 1 spatial dimension, 1 time dimension partial differential equation utility, "pdepe". The user needs to supply 3 functions; the differential equation, the initial conditions, and the boundary conditions. As an alternative to the home built grid solution, this package is versatile and useful. Indeed, it will be used heavily in the later section on quantum mechanics. Different initial heat pulse shapes can be chosen and their time development explored. The value of the conductivity can also be varied. The conductivity relates the size of the system to the time development, since reading the equation says that the conductivity, σ, has dimensions of length squared divided by time.

$$\frac{\mathrm{d}^2}{\mathrm{d}x^2}T = \left(\frac{1}{\sigma}\right)\frac{\mathrm{d}}{\mathrm{d}t}T \tag{4.15}$$

The user chooses the conductivity and a few initial temperature distributions. Other shapes could be tried by editing the code appropriately.

```
% Look at heat diffusion from an initial heat pulse
% numerical pde in 1 dimension, MATLAB utility used
% boundary and initial conditions
% Heat Pulse, U(x,t) = 0 at x = (-L,L) and U(x,0) = U(x),
x = (-ab,ab)
% dU/dt = a*d2U/dx2, a in m^2/sec
Itype = 3;  % 1 = Constant, 2 = x^2, 3 = cosx, 4 = x);
tau = ab .^2 ./(aa .*pi .*pi) % typical time(sec) given
conductivity and typical length
```

```
tau = 0.0188
```

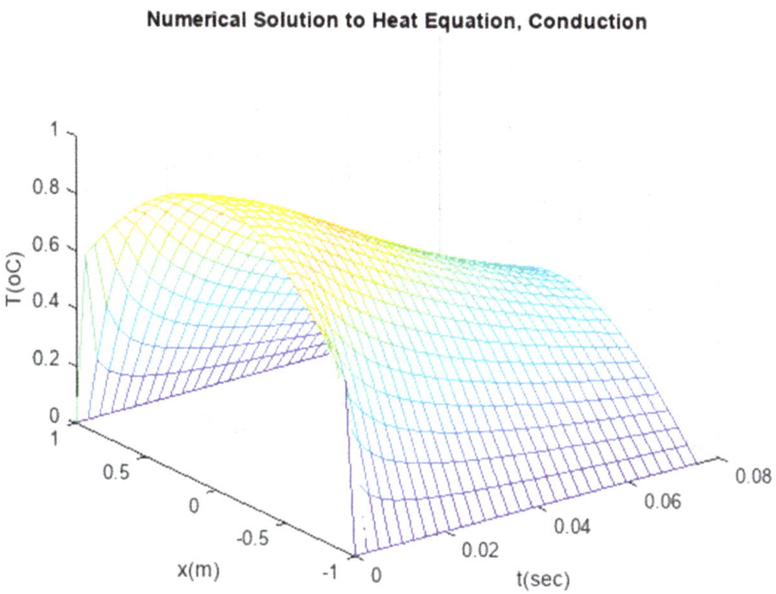

Figure 4.19: Time evolution of an initial temperature distribution as a function of position.

4.15 Airfoils

The solutions for flow past a sphere are well known and appear in the code for terminal velocity in the case of Stokes equation. In order to model an airfoil one can us complex variables to make a circle into an "airfoil", which here is a deformed circle pulled to a point at a location on the circumference. The transformed shape can be immersed in a constant velocity field which is rotated to an "angle of attack" by specifying u and v of the constant field. Analytic functions of a complex variable satisfy the Laplace equation, and can be used in many ways by choosing the appropriate transformation. The plots are made on a grid. The user chooses the angle of attack. The contours of the potential are displayed and the "gradient" utility is used to display the "lift" of the airfoil.

```
% Flow_Airfoil - incompressible and irrotational flow 2-D -
stationary streamlines and velocity
% Joukowsky profiles using complex variables and mapping
% Initialize parameters, x,y grid, and potential psi
% real part of psi is velocity potential, imag part is stream
function, v is del(psi)
%
kkk = exp(-j .*alf); % parameter for angle of attack, rotation of
incident flow
eta = z  + sqrt(z .^2 - 1.0) ; % map airfoil to off center circle -
root set by condition psi ~ a*z at large z
p = (kkk .*(eta + ra)) ./(1.0 + ra); % map off axis circle, radius
1+a to unit circle
psi = (p + 1.0 ./p); % map unit circle to uniform flow, slit
z = +1 to -1
```

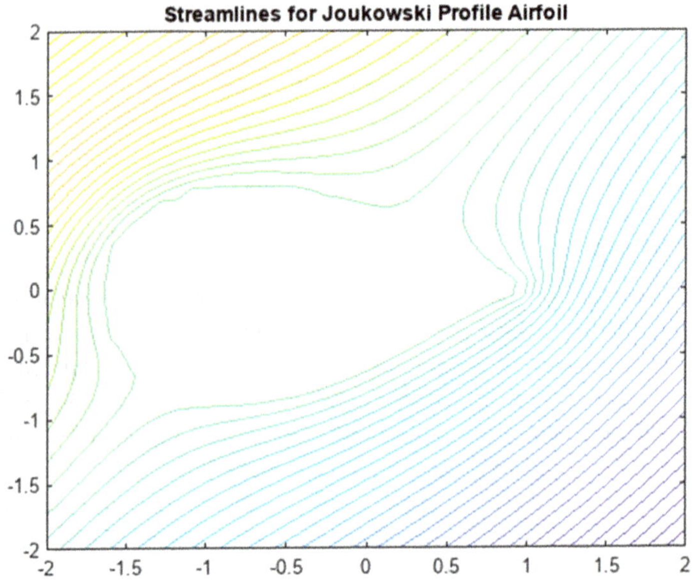

Figure 4.20: Streamlines for an airfoil model with a user chosen angle of attack.

```
% use gradient for the velocity computation, "lift" is evident
```

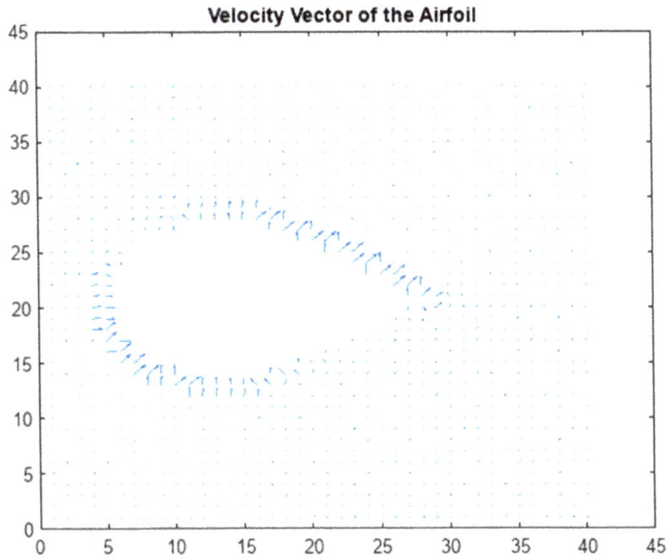

Figure 4.21: Velocity vectors for an airfoil model illustrating the "lift" of a specific angle of attack.

4.16 The Tides

The action of the tides is well known. It was Newton who first calculated the tidal height h and fully understood the physics. Indeed, tidal forces are local and intrinsic to gravity, as will be seen in a later discussion of General Relativity (GR) whereas the main acceleration, g, can be made zero by going to a free falling reference frame and so a constant acceleration cannot be intrinsic to gravity. For a location on Earth at a height z above the surface along an axis to the location of the Moon a distance r_o away with x transverse to z, the forces on an extended body are:

$$F_z = -\mathrm{GM}[1/(z + r_o)^2 - r_o^2] \sim 2\mathrm{zGM}/r_o^3, \quad F_x \sim \mathrm{xGM}/r_o^3 \qquad (4.16)$$

These "tidal" forces tend to elongate an extended body longitudinally and compress the body transversely and they scale as the distance from the center of a body divided by the cube of the distance from the attracting body. The forces can be derived from a tidal potential $V(x, y)$ which can also be expressed in polar coordinates

$$\begin{aligned} V(x, z) &= -\mathrm{GM}/r_o^3[z^2 - (x^2 + y^2)/2], \\ V(r, \theta) &= -(\mathrm{GM}/2r_o^3)[(3\cos^3\theta - 1)r^2] \end{aligned} \tag{4.17}$$

The tidal height difference is then, $\Delta h = (3\mathrm{GMr}^2/2gr_o^3)$ which is about 0.6 m for the Earth-Moon system. A schematic movie of the effect of the tides illustrates both the longitudinal and transverse effects. In the discussions of general relativistic (GR) effects the tidal nature will become obvious. Indeed, the "antenna" used to detect gravitational waves was deformed by the tidal forces carried by the metrical distortions of the gravity waves. That discussion is deferred for now. However, it is the tidal effects which are central to GR and not an acceleration like g.

```
% Tides due to the Moon attraction on the oceans - twice per day
```

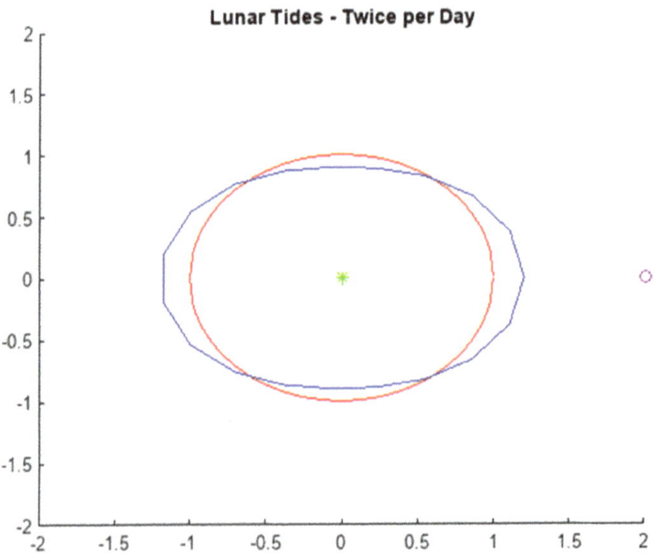

Figure 4.22: Schematic view of tidal effect of the Moon on the Earth. The red circle represents the Earth, while the blue line shows the tidal response of the oceans.

Chapter 5

Waves and Optics

"Time flows away like the water in the river." Confucius

"Descartes constructed as noble a road of science, from the point at which he found geometry to that to which he carried it, as Newton himself did after him. ... He carried this spirit of geometry and invention into optics, which under him became a completely new art." Voltaire

5.1 Mirror Aberrations

Begin with geometric optics. Light rays go in straight lines and the angle of incidence on a mirror equals the angle of reflection. In this approximation optics is geometry. Mirrors are widely used to focus an image. However, depending on the shape of the mirror, the focal point may not be the same for a spatial spread of incident parallel rays. Explore focal point aberration in spherical and parabolic mirrors. Find the angle of incidence of a parallel beam of light onto the mirror surface. Then construct the angle of reflection and the intersection of that ray with the mirror axis.

First the spherical mirror, easy to grind but with large aberrations. Only y values approximately <0.2 with respect to the focal point scale of ~ 0.5 have a reasonably tight bundle of focused rays. Then the parabolic mirror, harder to grind but with much smaller image aberrations. Note the large y range of the parabolic mirror compared to the spherical mirror focal point scale of ~ 2.0 and the much smaller, but not strictly zero spread of the focal point.

Astronomers want large area mirrors to see further out in space and parabolic mirrors are the shape of choice. The code is just straight lines in 2 dimensions. The mirror surface is red, the incident and reflected rays are blue, and for the parabolic mirror the bisector is in green.

```
% Ray Trace the Focus of a Spherical Mirror and Look at Aberration
% Focal Point, R/2, of a Spherical Mirror, Radius R = 1
% Incident parallel rays from the left (x < 0)
ymax = 1; % Fill the mirror
```

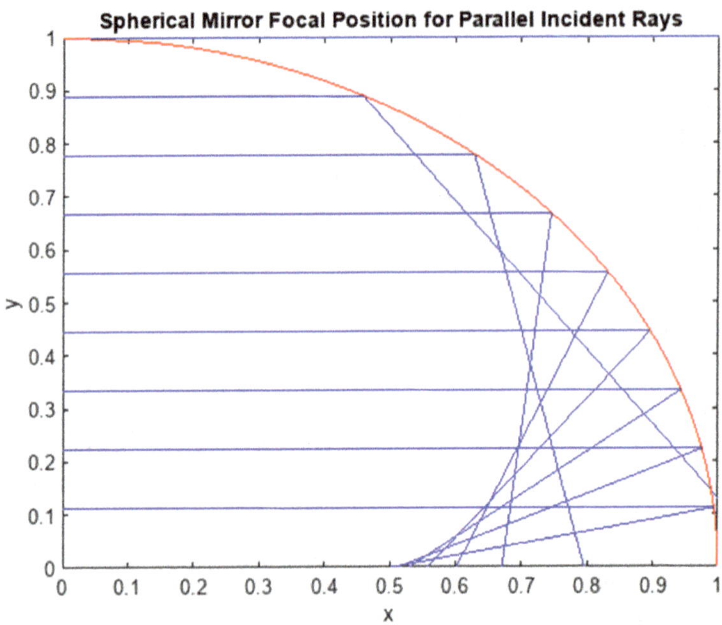

Figure 5.1: Ray trace for parallel rays incident on a spherical mirror.

```
%  now look at paraboic mirror - no speherical abberations
a = 2; % focus at (a,0)
% parallel beam in - intersect parabola at x
x = (y .*y) ./(4 .*a);
% angle of incidence, normal = [1 , -y/2a]
```

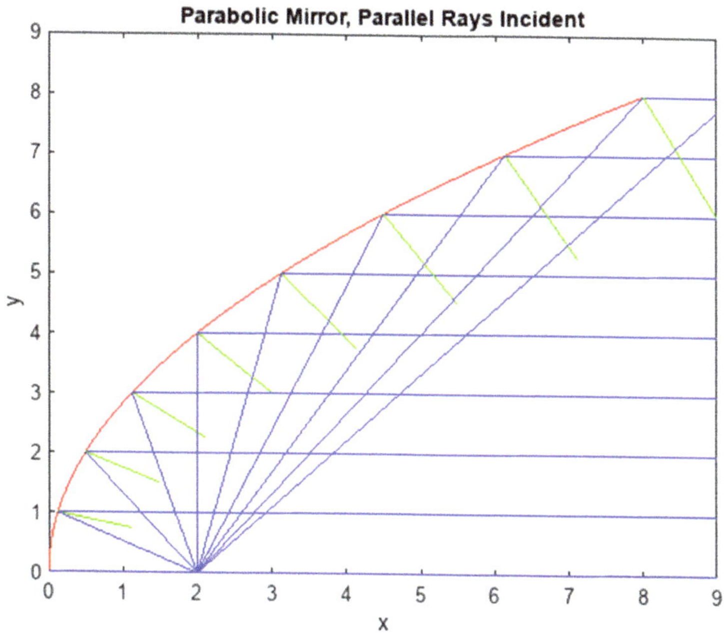

Figure 5.2: Ray trace for parallel rays incident on a parabolic mirror.

5.2 Optical Interface

In many situations in optics a wave encounters an interface where the index of refraction of the medium changes abruptly. Indeed, the same situation obtains in quantum mechanics, as will be discussed later. After encountering the interface there will be, in addition to the incident wave train, a reflected wave and a transmitted wave. The intensity of these 3 waves is found by requiring continuity of the wave itself and the wave vectors, k. In quantum mechanics the continuity of the wave function and the derivative of the wave function is understood to be a requirement that probability and momentum is conserved at a 1 dimensional interface.

```
% set up an interface between plane waves, wave vectors
% k1 for x <0 and k2 for x > 0, (exp(i(kx-wt))
% for Quantum V = 0  for x < 0 and Vo for x > 0
% E = hbar^2k1^2/2m x < 0, (E-vo) = hbar^2k2^2 x > 0
% wave incident from the left, I, with reflected wave R
% and transmitted wave T, take I = 1, Vo could be < E or > E
```

The wave and the first derivative (wave momentum) should be continuous at the boundary. For normal incidence with incident, reflected and transmitted waves I, R and T.

$$\begin{aligned} 1 + R &= T \\ k_1(1 - R) &= k_2 T \end{aligned} \tag{5.1}$$

```
syms k1 k2 R T s
eqns = [1+R-T == 0  k1*(1-R)-k2*T == 0 ];
s = solve(eqns, [ R T])
```

```
s = struct with fields:
    R: (k1 - k2)/(k1 + k2)
    T: (2*k1)/(k1 + k2)
```

For incidence at an angle, the 2 dimensional problem is treated by requiring that the angle of incidence is equal to the angle of reflection, The conservation of momentum components at the interface implies that:

$$k_1 \sin\theta_1 = k_2 \sin\theta_2 \tag{5.2}$$

which is Snell's law.

5.3 A Longitudinal Slinky

Construct a model of ions bound in a lattice. Assume that they have only short range forces and therefore interact only with nearest neighbors in a way approximated as a spring. Ignore transverse motion and explore only a longitudinal degree of freedom. Look at the motion of the 8 mass points and 7 springs with a simple single initial displacement of the fourth spring. The waves are quite complex and similar to those of the "Slinky" toy of years past. The user can initiate both positive and negative displacements and view the resulting "movie" of the resulting waveform. The user selects the amplitude of the displacement, positive or negative of the fourth spring.

```
% look at linear array of 7 1-d springs, longitudinal, 8 mass points
k_m =  1.0 ; % spring constants
% mass points at ends, # 1 and # 8, 8 masses, 7 springs
dx1 = -0.019; % initial displacement of the fourth spring
tscale = 1.0 ./sqrt(k_m); % time scale set by individual spring
frequencies
```

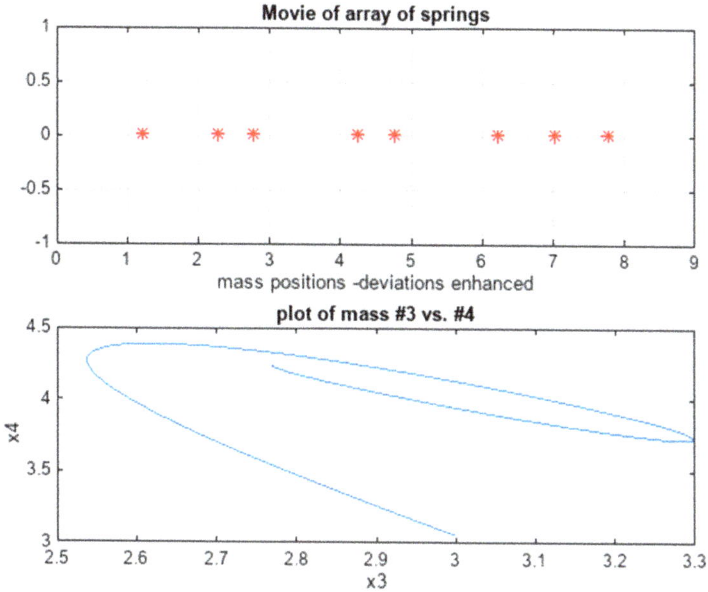

Figure 5.3: Model of a crystal as a series of masses on nearest neighbor springs. Top: snapshot of the position of the 8 "ions". Bottom: trajectory of springs 3 and 4.

```
% coupled spring equations
% end points are fixed.
% points 2-6 have 2 nneighbors, left, right
```

5.4 Beat Frequencies

Waves can be added with different amplitudes and phases. They overlap in such a fashion that there are "beat" frequencies which depend on the differences between the frequencies of the 2 waves as well as the sums. The user can first study this effect by varying the ratio of the amplitudes of the 2 waves. The total intensity, I_{12}, depends on the amplitude ratio and the phase difference, $\Delta\phi = (\omega_1 - \omega_2)t + (\phi_1 - \phi_2)$.

$$\begin{aligned} A_{12} &= a_1 \cos(\omega_1 t + \phi_1) + a_2 \cos(\omega_2 t + \phi_2) \\ I_{12} &= |A_{12|}|^2 \end{aligned} \tag{5.3}$$

In the simplest case of equal amplitude, $a_1 = a_2$, and zero initial phases $\phi_1 = \phi_2$ the trig identity:

$$\cos(\omega_1 t) + \cos(\omega_2 t) = 2\cos((\omega_1 + \omega_2)t/2)\cos((\omega_1 - \omega_2)t/2) \tag{5.4}$$

shows that there is an envelope with the frequency sum and a "beat" factor with the frequency difference. In other cases the result for I_{12} is not so transparent. The user first selects the amplitude ratio using an "Edit Field". Following that the amplitudes are fixed to be equal, but the frequencies are taken to be different.

```
% adding 2 waves - beats
% Harmonic Oscillation
% T = 1/f,  w = 2*pi*f, Period, Frequency;
%  d2x/dt2 = kx/m, wo^2 = k/m
% x = Acos(wot)
% Examples: pendulum wo^2 = g/l, circuit = 1/L*C
% Wave Number = k = 2*pi/lambda = w/c, v = f*lambda
% Two Waves phase difference = path difference*2*pi/lambda
a1_a2 = 2 % Ratio of the Amplitudes
```

```
a1_a2 = 2
```

```
% span the phase differences
I12 = ((a1_a2) .^2 + 1.0 + 2 .*a1_a2 .*cos(delphi));
```

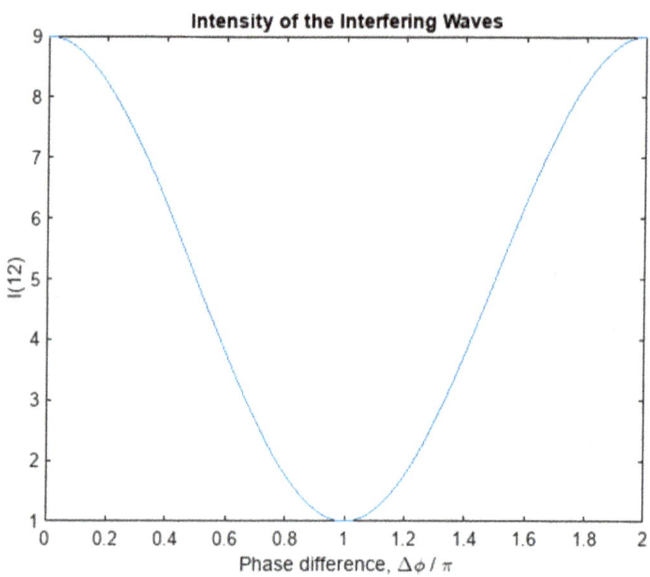

Figure 5.4: Intensity of the addition of 2 waves with differing phases as a function of the phase difference.

Next, the result of having 2 waves with equal amplitude but with different individual frequencies is displayed.

```
% Waves of Equal Amplitude, = 1, and Phase
% Fractional Difference of the 2 Frequencies, r = (w1-w2)/(w1+w2)
% factor cos((w1+w2)*t/2) * factor cos((w1-w2)*t/2),
w = (w1+w2)/2 - beats
%
```

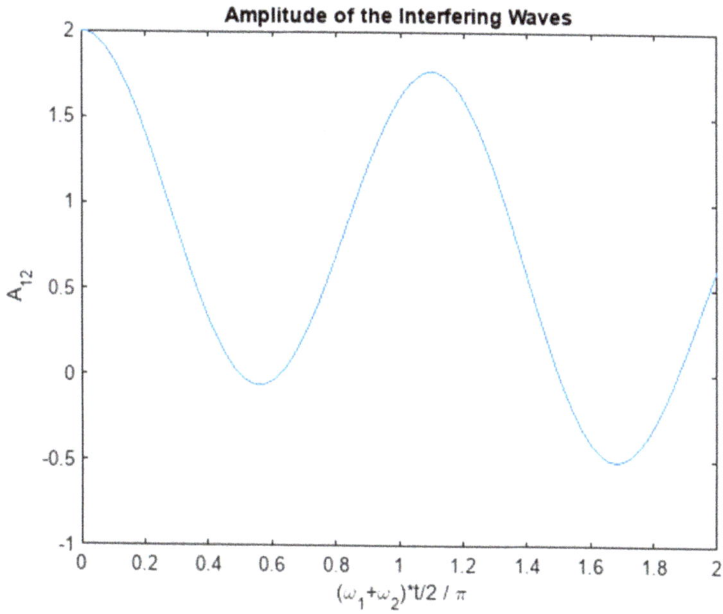

Figure 5.5: Amplitude for the addition of 2 waves with different frequencies as a function of time.

Finally, the ratio r is varied and a "movie' is made of the full waveform for r varying from 0 to 20. At large r values, there is an envelope with slow amplitude modulation which encloses a faster oscillation. As usual with a "movie" there is a playback possibility.

```
ratio = linspace(0,20,20);
AA(i,:) = 2.0 .*cos(wt(:)) .*cos(ratio(i) .*wt(:));
```

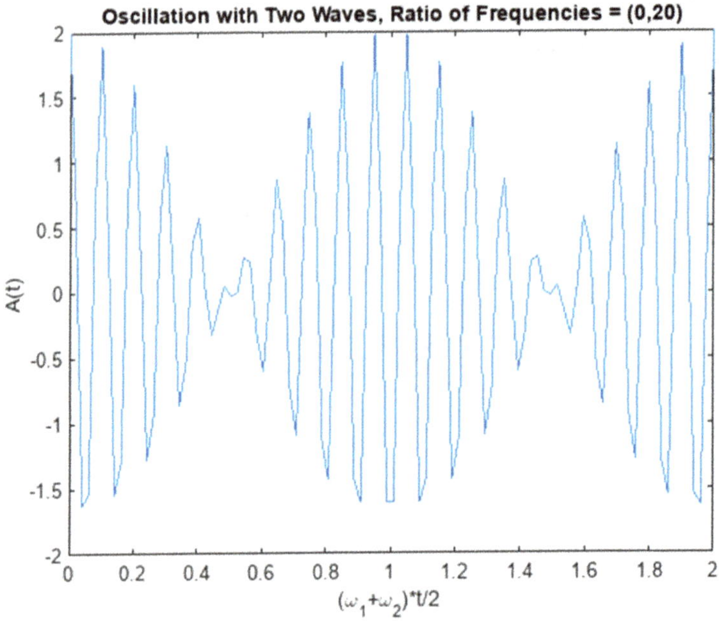

Figure 5.6: Last frame of a movie of the amplitude for the addition of 2 waves with different frequencies as a function of time.

5.5 Lissajous Figures

Another summation of waves, in this case with different phases and frequency ratios is familiar to devotees of old science fiction movies. In this case the amplitude ratio of the 2 waves is fixed, but the phase difference and frequency ratio can be chosen using the "sliders" provided. One wave sets the x axis value, the other sets the y value (think of oscilloscope inputs). If the phase difference is small a simple y ~ cos(x) waveform is the result. The user sets the phase difference and the frequency ratio of the x and y axes.

```
% Look at a 2-d oscillator, Lissajous figures
% Two Dimensional Oscillator, Different Amplitude, Phase and
Frequency
a =  2; % Amplitude Ratio y/x
phi = 75; % Phase Difference y - x, degrees
wxy = 4.4; % Frequency Ratio y/x
wxt = linspace(1,4 .*pi,100);     % wx = 1, 2 periods
x = cos(wxt);
y = a .*cos(wxy .*wxt + phi); % second oscillation
xmax = max(x); xmin = min(x); ymax = max(y); ymin = min(y);
```

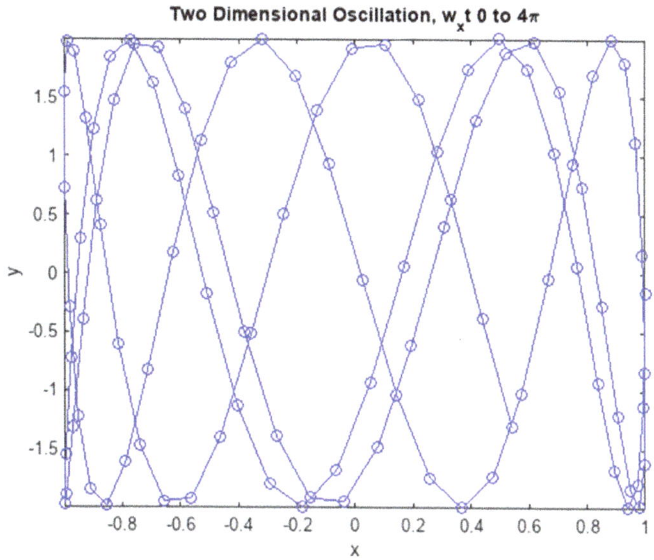

Figure 5.7: Plot of (x, y) for a user chosen choice generating the Lissajous waveform.

5.6 Drums

A string in 1 dimension, or a drum in 2, has a distinct set of allowed frequencies. They are found by imposing boundary conditions on the vibrating object at locations where the string or membrane is fixed. The solutions are found using the standard method of separation of variables, assuming a product solution in the relevant dimensions. For a rectangular drum the sin and cos functions with discrete frequencies satisfy the boundary conditions of vanishing on the x and y boundaries, while for a circular drum the radial zeros of the Bessel function solutions are employed. In the circular case the wave equation for the amplitude of the drum membrane M is:

$$\begin{aligned} \frac{\partial^2}{\partial t^2}M &= v^2\left(\frac{\partial^2}{\partial x^2}M + \frac{\partial^2}{\partial y^2}M\right) \\ M_{m.n}(r,\phi,t) &= \cos(v\lambda_{m,n}t)J_m(\lambda_{m,n}r)\cos(m\phi) \\ J_m(\lambda_{m,n}(a)) &= 0 \end{aligned} \tag{5.5}$$

The azimuthal functions are limited to cos as is the time dependence, for simplicity. The actual solution can be any superposition of these normal modes. In the code the radius is set to 1 as is the velocity v. The boundary conditions are radial, imposed at r = a. Details, as usual, are found in the "comments - % - in the body of the code. Both rectangular and circular drums are explored. The first few numeric roots of the Bessel function are tabulated in the code. The user chooses m and n over a range, m = 0, 1, 2 and n = 1, 2, 3.

```
% Rectangular and circular Drum - 2d Wave Eq
% solve by separation of variables
% Rectangular Drum of x Width a = 1 , y width b
% Spatial: sin(m*pi*x/a)sin(n*pi*y/b)
% Temporal: cos(2*pi*vn,m*t), vm,n = 1/2*sqrt(T/rho)
% Temporal: sqrt((m/a)^2+(n/b)^2)
n = 2; m = 3;
%  x mode m, y mode n
```

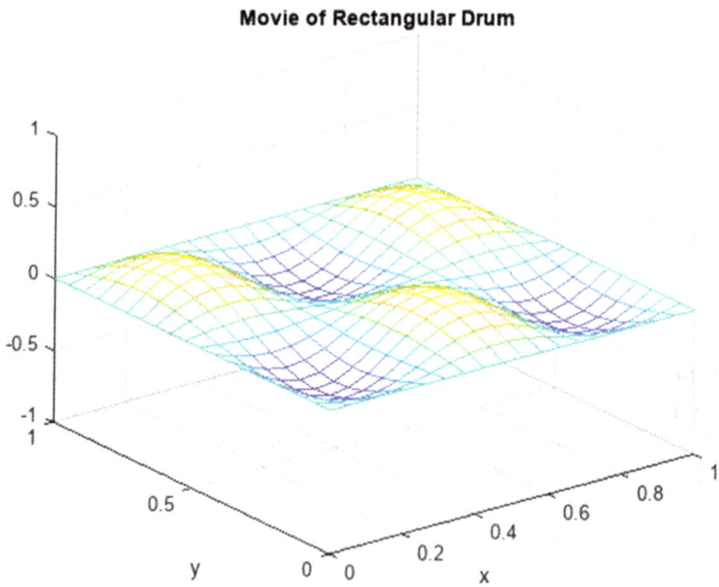

Figure 5.8: Last frame of a movie of the drum amplitude for a rectangular drum.

```
% now the circular drum
% Circular Drum of Radius r = 1
% Radial Jm(Lmnr), Azimuthal -  C*cos(m*phi)+D*sin(m*phi) Initial
conditions, D = 0
% Temporal, cos(cLmnt) sin term = 0 - IC, speed c = 1, J of type m,
with root Lmn
m = 1
```

```
m = 1
```

```
n = 1
```

```
n = 1
```

```
% m = 0,1,2,  n = 1,2,3;
roots = [2.405, 5.520,8.654;3.817,7.016,10.17;5.136,8.417,11.62];
% Bessel Function Jm, m = 0,1,2 Order of the Bessel Root n = 1,2,3
```

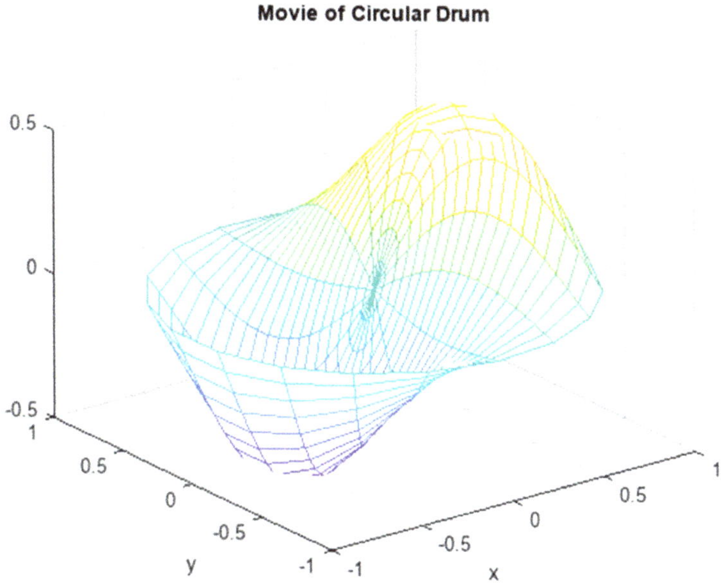

Figure 5.9: Last frame of a movie of the drum amplitude for s circular drum.

5.7 Diffraction at Apertures, Near and Far Zone

Diffraction at an aperture is a fundamental property of wave phenomena. Matter waves diffract in addition, as noted in the classical double slit experiment of quantum mechanics. For these problems MATLAB supplies a suite of special functions which are needed to quantitatively describe the diffractive phenomena. In 2 dimensions the Bessel functions are needed. For the single slit, secondary maxima are observed. These first patterns are evaluated in the "far zone", far from the diffracting apertures on the scale of the wavelength, $\lambda \gg d$. More correctly, the Fraunhofer far zone, with d a source size which is $\gg \lambda$, refers to distances r, such that $4\lambda r > d^2$, A "slider" is used to set the slit width. For a single slit of width d in the far zone, the phase difference for a wave with wave vector k and wavelength λ is $\alpha = \pi \sin\theta(d/\lambda) = (kd\sin(\theta)/2)$ and the intensity I is $I = (\sin\alpha/a)^2$. The results for a circular aperture of diameter D are similar to those for a single slit, $I = [2J_1(\alpha)/\alpha]^2$, $\alpha = \pi D\sin(\theta)/\lambda$.

```
% First Franhoufer Diffraction at a Single and
Double Slit 1-d and circular slit;
% Input Ratio of Slit Width the Incident Wavelength ~ Resolving
Power
% Path Difference = d*sin(theta), Phase Difference =
pi*d*sin(theta)/lambda
I1d = (sin(dphi) ./(dphi)) .^2; % single 1-d slit
% Phase Difference for Circular Aperture, k*D*sin(theta)/lambda,
D = diameter
% D = dslit, lambda = 1
Icirc = (2.0 .*besselj(1,dphi) ./dphi) .^2 ;
```

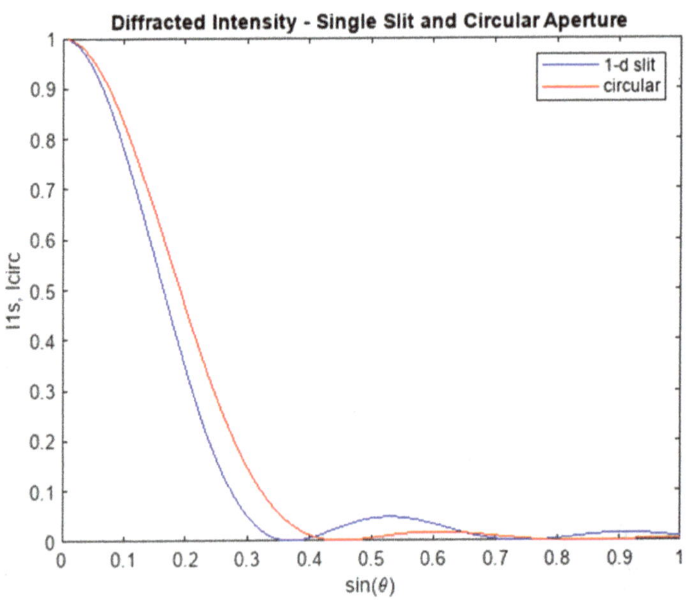

Figure 5.10: Intensity for diffraction from a slit and a circular aperture as a function of observation angle.

Following the single slit, the diffraction from a double slit can be explored. The important parameter is the 2 slit spatial separation compared to the wavelength of the waves. This ratio can be chosen by the user and the resulting shapes of the diffracted intensity explored. There is an envelope of secondary maxima but now modulated by the 2 slit oscillations. The patterns are determined using the "slider" to set the 2 slit separation relative to the wavelength.

```
% Input Ratio of Two Slit Separation to the Incident wavelength
al = 11.4;phase = pi .*sinth .*al;
I2s1d = I1d .*((cos(phase)) .^2);
```

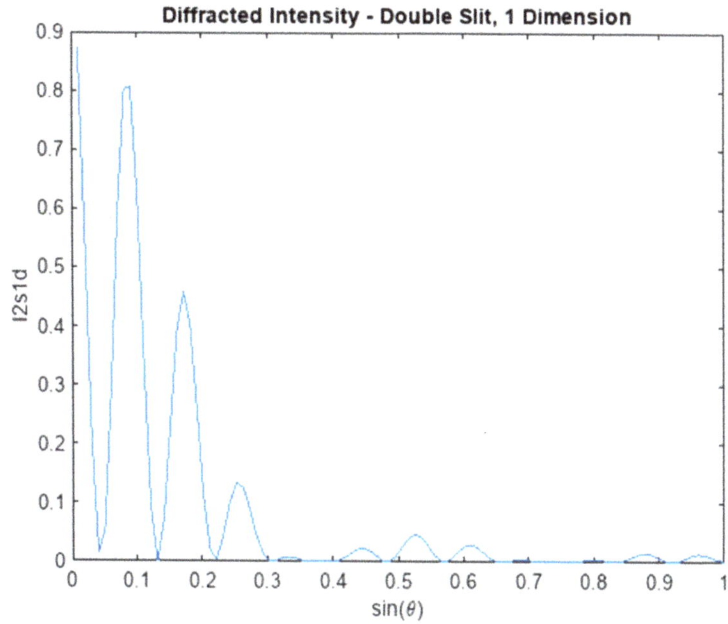

Figure 5.11: Intensity for diffraction from a double slit as a function of observation angle.

Returning to diffraction in a single dimension, one can also look at the situation when the observation is not in the "far zone". There are 3 geometries which are possible as commented below by way of explanation. In this case the situation in the "near zone" is displayed. The special functions that are needed are called the "Fresnel Integrals". Indeed MATLAB has a library of special functions for almost every situation in classical physics, Three types of apertures are possible choices. In the case of the long slit the user can choose the width h.

```
% Itype = 1, Absorbing Screen Covers Lower 1/2 Plane
% Itype = 2, Long Slit of Width h
% IType = 3, Square Aperture of Width h
% Observe at zo Distance Behind Screen/Slit/Aperture
% Fresnel is Near Zone with zo ~ lambda
% zo in Wavelength Units
zo = 5.0;
% variable is vertical distance y in units of wavelength
S = fresnels(w);
I = (C + 0.5) .^2 + (S + 0.5) .^2;
```

```
% Slit Width in Wavelength Units, Itype == 2
w1 = (yo - h ./2.0) .*sqrt(2.0 ./zo);
w2 = (yo + h ./2.0) .*sqrt(2.0 ./zo);
C1 = fresnelc(w1);
C2 = fresnelc(w2);
S1 = fresnels(w1);
S2 = fresnels(w2);
I = ((C1 - C2) .^2 + (S1 - S2) .^2) ./2.0;
% Itype = 3, Square Slit Width in Wavelength Units
w1 = (yo - h ./2.0) .*sqrt(2.0 ./zo);
w2 = (yo + h ./2.0) .*sqrt(2.0 ./zo);
v1 = (xo - h ./2.0) .*sqrt(2.0 ./zo);
v2 = (xo + h ./2.0) .*sqrt(2.0 ./zo);
```

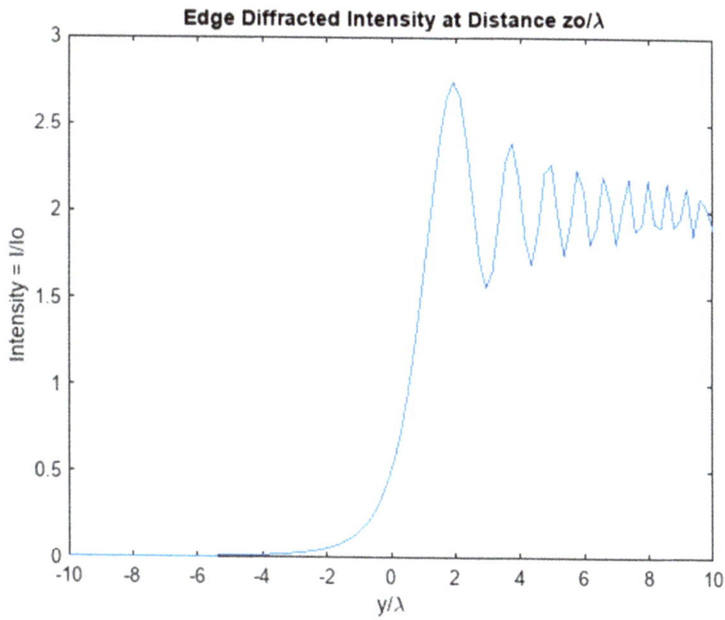

Figure 5.12: Intensity for diffraction from an edge as a function of distance from the edge in the near zone.

Chapter 6

Quantum Mechanics

"Those who are not shocked when they first come across quantum theory cannot possibly have understood it." Niels Bohr

"Quantum physics thus reveals a basic oneness of the universe." Erwin Schrodinger

This section opens the examples of physics in the twentieth century, as opposed to the previous 5 sections covering classical mathematics and physics. In quantum mechanics Nature has a fundamentally minimum structure characterized by the constant $\hbar$ which has dimensions energy times time or position times momentum. That sets a basic scale. The number of possible states of a system is limited by a fundamental cell size in position — momentum space $\mathrm{dx\,dy\,dz\,dp}_x\mathrm{dp}_y\mathrm{dp}_z \sim \hbar^3$ which was previously defined to be "phase space".

An intermediate step on the way to true quantum mechanics was a Bohr type analysis of a bound state.

6.1 Energy Levels, System Size and Force Laws

Scattering experiments can be used to infer the forces between the projectile and the target. Another technique is to look at the energy levels of a system. They are also defined by the forces between the constituents of a system and can be measured by observing the photons emitted during transitions between those energy levels. The hydrogen atom has energy levels $\sim 1/n^2$ for a potential $\sim 1/r$. The SHO has energy levels $\sim n$ with a potential $\sim r^2$. Going back to the original Bohr atom, one can require that the deBroglie wavelength, $\lambda_{\mathrm{dB}} = 2\pi\hbar/p$ corresponding to an orbit of radius r contains an integer number of phase advances, so that the wave is self-sustaining, $\lambda = r/n$. This condition was imposed to keep the atoms from radiating away all their energy. The total energy is then minimized.

$$\begin{aligned} V &= \alpha/r^b \\ r &= (n^2/m\alpha b)^\gamma, \quad \gamma = 1/(2-b) \\ E &= n^{-2b\gamma} m^{b\gamma} \alpha^{2\gamma} \end{aligned} \tag{6.1}$$

For bound states, this "Bohr" style analysis of the radial quantum number, n, can be explored below. It works well for the hydrogen atom, both energy level and radius. The user can choose the exponent of the potential, V.

```
% Look at bound states, ground state of Schroedinger Eq - force
laws
% potential is assumed to be a power law.
% energy e, potential V = a/r^b, mass m , radial quantum number
% System Size as a Function of n, m, a and b
aa = (n^2/(m*a*b))^(1/(2-b));
% Energy as a Function of n, mass, coupling , and Power Law
e = n^((-2*b)/(2-b))*m^(b/(2-b))*a^(2/(2-b));
% Potential power: Coulomb = 1, SHO = -2
bb = 1;
% System Size
aaa = (n^2/(m*a*bb))^(1/(2-bb))
```

aaa = $\dfrac{n^2}{a\,m}$

```
c1=-(2.0 .*bb) ./(2.0 -bb); c2=bb ./(2.0 -bb); c3=2.0 ./(2.0 - bb);
% System Energy
ee=(n^c1)*(m^c2)*(a^c3);
ee
```

ee = $\dfrac{a^2 m}{n^2}$

6.2 Free Particle — 1d and 3d

The changes from classical to quantum physics are profound. Since matter has wave properties there is no point particle. Rather there is a wave function, ψ, which contains all the information possible and whose square is the probability to observe the particle. There is no velocity, but the spatial derivative indicates how rapidly the wave function varies with position.

That indicates that the momentum is $p = i\hbar\frac{\partial}{\partial x}\psi$ in 1 dimension. Similarly, $E = i\hbar\frac{\mathrm{d}}{\mathrm{d}t}\psi$ also has the correct dimensions. A wave equation in 1 dimension is then $(T+V)\psi = E\psi = -(\hbar^2/2m)\frac{\mathrm{d}^2}{\mathrm{d}x^2}\psi + V\psi = i\hbar\frac{\mathrm{d}}{\mathrm{d}t}\psi$. A free particle, $\mathrm{V} = 0$, then has plane wave solutions, $e^{i(\mathrm{kx}-\omega t)}$ where $p = \hbar k$, $E = \hbar\omega$ and $E = p^2/2m$.

The free particle also has a closed form solution in 3 dimensions. There is a "centrifugal potential" analogous to the classical result which was applied to reduce the Kepler problem to a 1-d problem by defining a repulsive potential $\sim L^2/r^2$. This term appears in the quantum problem from quantizing the angular motion and obtaining a radial term $\sim l(l+1)\hbar^2$. The solutions factorize into spherical harmonics times a radial wave function, $\psi(r,\theta,\phi) \sim u(r)Y_{l,m}(\theta,\phi)/r$. Since this is a 3d problem there are 3 quantum numbers, n for radial, l for polar, and m for azimuthal coordinates. The radial solutions of the radial wave equation for the Schrodinger equation, $\frac{\mathrm{d}^2}{\mathrm{d}r^2}u(r) - [l(l+1)/r^2]u(r) + q^2u(r) = 0$, $q^2 = 2\mathrm{mE}/\hbar^2$ are radial Bessel functions. The solutions show that higher angular momentum states, with larger l, are pushed out from the origin, as expected.

```
% 3-d Schroedinger Eq. for free particle, u
syms l u(r) q r uu Jo J1 ode C2
ode = diff(u,r,2)-u*l*(l+1)/r^2+q^2*u == 0;  % q^2=2m(E)/hbarc2,
psi=u/r*Ylm
u = dsolve(ode)
```

u = $C_1\sqrt{r}\ \mathrm{J}_{l+\frac{1}{2}}(qr) + C_2\sqrt{r}\mathrm{Y}_{l+\frac{1}{2}}(qr)$

```
% Y Bessel diverges at r = 0, C2 = 0, look at l =0 and l = 1
solutions
Jo = (besselj(1/2,q*r))
```

Jo = $\dfrac{\sqrt{2}\sin(qr)}{\sqrt{\pi}\sqrt{qr}}$

```
J1 = (besselj(3/2,q*r))
```

J1 = $-\dfrac{\sqrt{2}\left(\cos(qr) - \frac{\sin(qr)}{qr}\right)}{\sqrt{\pi}\sqrt{qr}}$

```
% J for l = 0, 1 , use Bessel explicit results
u = (besselj(l+1/2,q*r))^2*r;     % psi^2*r^2 - density = u^2
% numerical inputs
```

```
me = 511000.0;  % eV - electron mass in eV
hbarc = 2000.0;   % ev*A
E = 10.0;   % eV, T no V
k = sqrt(2.0 .*me .*E) ./hbarc ;
% Wave Number k for a 10 eV Free Particle
```

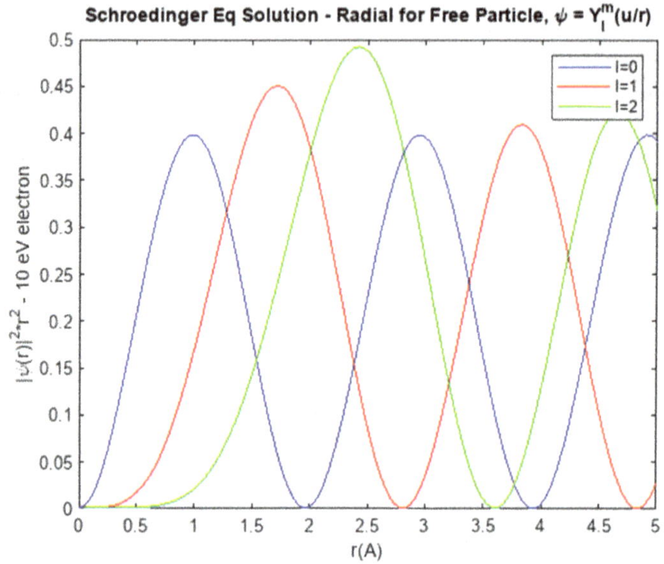

Figure 6.1: Radial wave functions for the lowest angular momentum quantum number for a free particle.

6.3 The Hydrogen Atom

Classical physics has a problem; stable atoms cannot exist. For example, hydrogen has an electron bound to a proton in some orbit. The electron is accelerated in that classical orbit, so it must radiate. If it radiates it loses energy and ultimately collapses to a point. However, stable atoms exist, Hydrogen is observed to have a ground state radius, $a_o = (\hbar c/m_e c^2)/\alpha$ and an energy, inferred from spectroscopy and other data, of $E_o = e^2/8\pi\varepsilon_o a_o = -m_e c^2 \alpha^2/2$. The solution to this problem was the invention of quantum mechanics, where wave properties were ascribed to matter.

As in other fields of physics, there are few closed form solutions in quantum mechanics. The free particle solution was already defined for 1 and 3 dimensions. Several others will be explored in what follows. The hydrogen atom has a closed form solution in quantum mechanics. Indeed, it was one the initial successes of the "new physics". The energy level scales as $1/n^2$ where n is the principle quantum number, defining the radial dependence where n is any integer 1 and above. The mean radius scales as n^2. As the atom goes to a higher energy state, less bound, it's radius increases. The probability density has a number of maxima that scale the same way as the mean radius. The radial densities for $n = 1, 2$, and 3 are plotted below. The number of maxima is n, The angular solutions are defined later.

```
% Solutions for the Hydrogen Atom
% There are few Solvable Problems in 3-d QM. The hydrogen atom is
one
mec2 = 511000.0;  % eV - electron mass
hbarc = 2000.0;   % ev*A
% Bohr radius
labe = hbarc/mec2; % Compton radius
alf = 1.0 ./137; % fine structure constant
ao = labe ./alf;    % <R> ~ AO * N^2
Eo = -(mec2 .*alf .*alf) ./2.0;    % ground state energy
E1 = Eo ./4; E2 = Eo ./9; % 1/n^2 level spacing
ro = (3.0 .*ao) ./2.0; % ground state mean radius - mean radii in
ao units
r1 = ro .*4; r2 = ro .*9;
```

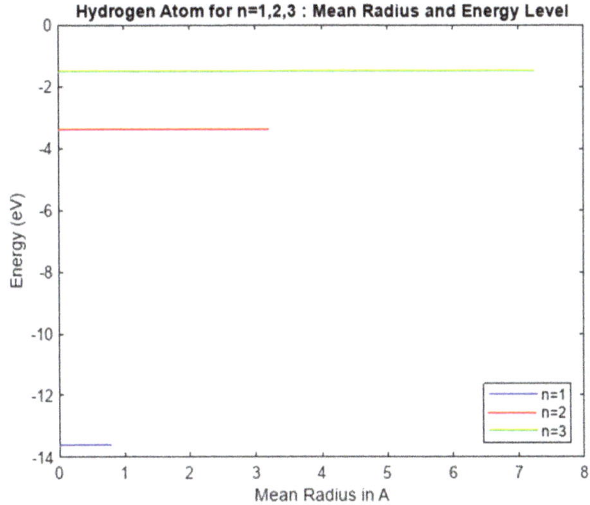

Figure 6.2: Bound state energy for the lowest radial quantum numbers for the hydrogen atom. The units used are Angstroms for historical reasons. 1 A = 0.1 nm.

```
% radius in Angstroms
q = r ./ao;
psi0 = (exp(-q)) .^2;
psi1 = ((exp(-q ./2.0) .*(1 - q ./2.0))) .^2;
psi2 = (exp(-q ./3.0) .*(1 - (2.0 .*q) ./3.0 + (2.0 .*q .*q) ./27.0)) .^2;;
```

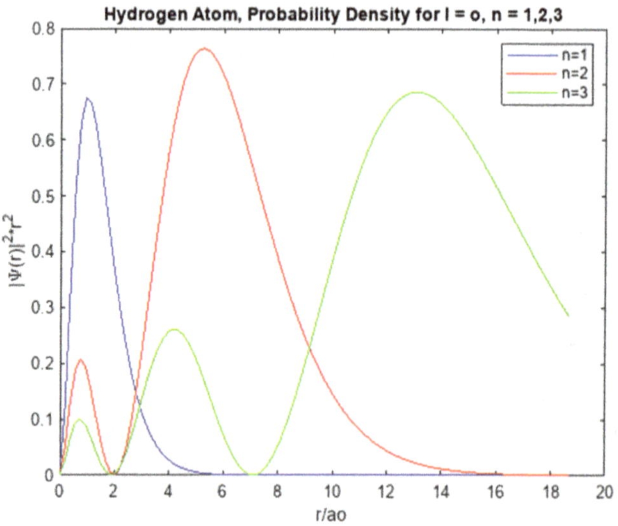

Figure 6.3: Bound state wave functions for the lowest radial quantum numbers for the hydrogen atom.

6.4 Hydrogen Angular Solutions

For a 3 dimensional problem there will be 3 quantum numbers needed to specify the state of the system. The energy is set by the radial, or principal, quantum number. The polar angle, θ, is associated with the classical angular momentum and has a quantum number ℓ with integer values from 0 to $n - 1$. The azimuthal coordinate has a quantum number m with values from $-\ell$ to ℓ. The solutions are called spherical harmonics $Y_m^\ell(\theta, \phi)$. In general the number of angular lobes increases with ℓ and at a fixed ℓ increases with the m value.

```
% QM spherical coordinates. Angular Ylm,plot 3-d
% now look at 3-d surface of angular dependence
% theta and phi locate the point direction - r the length
```

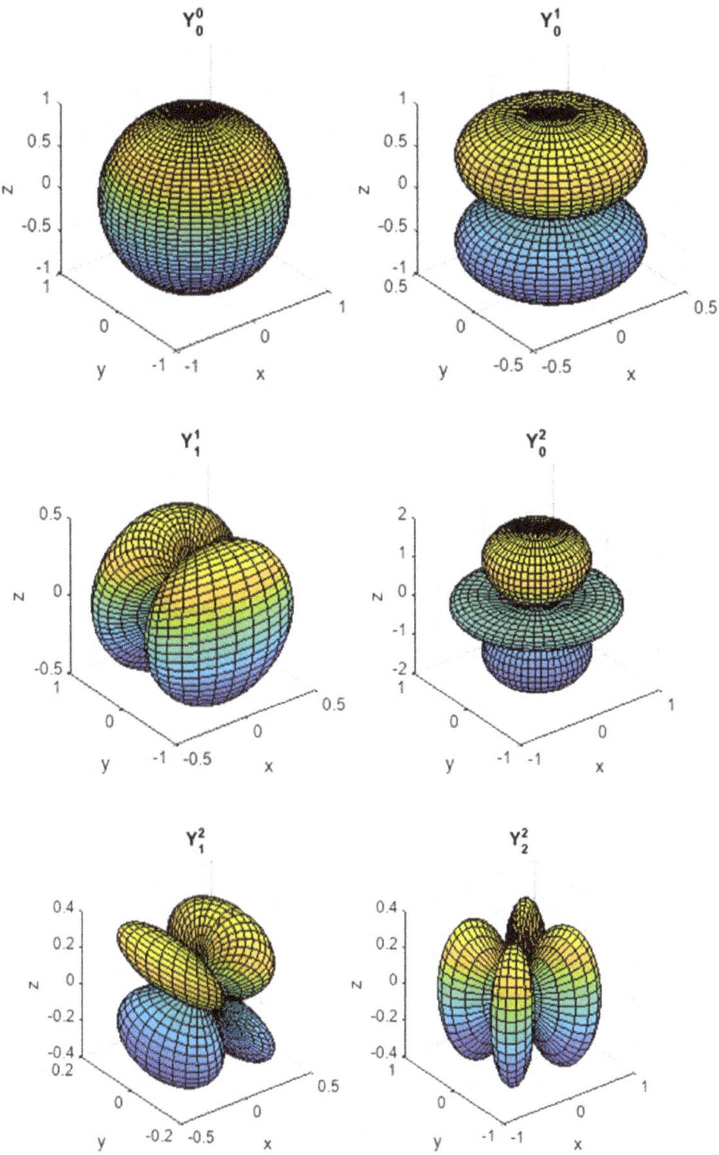

Figure 6.4: Wave functions for the lowest polar and azimuthal angle quantum numbers $\ell = 0, 1, 2$ and m running from $-\ell$ to ℓ.

6.5 Atoms and the Periodic Table

The simplest atom is the hydrogen atom. Solving the Schrodinger equation the solutions for the principal quantum number n and the effects of the centrifugal force as determined by the quantized angular momentum ℓ are $\psi(r, l) \sim r^{\ell}e^{-r/\mathrm{n}a_o}$ where the ground state radius, the Bohr radius, is $a_o = \hbar/\,\mathrm{mc}\,\alpha$. The radius is the Compton wavelength greatly increased by the inverse of α, which means the outer electrons of an atom are fairly loosely bound. The ground state energy is numerically $-13.6\,\mathrm{eV}$ and the ground state radius is $0.53\,\mathrm{A} = 0.053\,\mathrm{nm}$. An estimate of the orbital "velocity" is α which validates the use of NR quantum mechanics. The power law behavior of r with ℓ shows the centrifugal effect of the angular momentum, familiar from classical Newtonian orbits.

$$E_o = -m_e c^2(\alpha^2/2) \tag{6.2}$$

An understanding of the Fermi Exclusion Principle and the hydrogen atom solution allows an understanding of the periodic table of the elements. No 2 electrons can be in the same state. The spin 1/2 electron has 2 possible spin states. The 3 coordinates have 3 associated spatial quantum numbers. The angular momentum L has integer ℓ and integer z projection m. The number of possible states defined by n, since $n > \ell$ is then $(2\ell + 1)2$. For historical reasons $\ell = 0, 1, 2$ states are called s, p and d.

Using this information, one can look at the electronic structure of atoms and understand much of the classification of elements in the periodic table. If one ignores the Z electrons, the energy of any one electron might be estimated to be $E \sim E_o Z/n^2$ so that the innermost electrons would be pulled in close by the Z protons. Likewise the radius of an inner electron might be $\sim a_o/Z$. But for an outer electron the Z protons would be "screened" by the inner electrons so one might expect all atomic sizes to be comparable to that of hydrogen. The filling sequence would be $n = 1, 2$ states H and He, followed by 2 states with $n = 2$ (starting with Li), $\ell = 0$ and then 6 states with $\ell = 1$, ending with F and Ne. The next shell would be for $n = 3$, with 2 in $\ell = 0$, Na, 6 in $\ell = 1$ (ending in Cl and Ar) and 10 in $\ell = 2$ (starting with P and ending with Br and Kr. A sequence labelled by n starts with a metal having a loosely bound electron, and ends with a highly reactive base, with a need for an electron to complete the shell and then a noble gas with a closed n shell; a very unreactive element.

These approximations can be explicitly made and compared to the data from the periodic table. The ionization potential and the atomic size are both experimentally determined. In this way, chemistry is approximately understood.

```
% radius and ionization potential of atoms Z = [1, 18]
% For Hydrogen Atom, Eo = -13.6 eV = mec^2alpha^2/2
% For Hydrogen Atom, ao = 0.54 A = lambdabar/alpha, lambdabar =
hbar/mec = 0.004 A
% For Hydrogen Atom, Eo = e^2/2ao, beta = v/c = alpha
% data on ionization potential,energy needed to free 1 electron,
and radius
ion = [13.6 24.6 5.4 9.3 8.3 11.3 14.5 13.6 17.4 21.6 5.1 7.6 6.0
8.15 10.5 10.4 13.0 15.75];
rad = [0.37 0.5 1.52 1.11 0.88 0.77 0.70 0.66 0.64 0.70 1.86 1.60
1.43 1.17 1.10 1.04 0.99 0.94];
% Filling Sequence = (1s)^2(2s)^2(2p)^6(3s)^2(3p)^6,, low l filled
% first(smaller r, r^l), low n filled first(1/n^2)
```

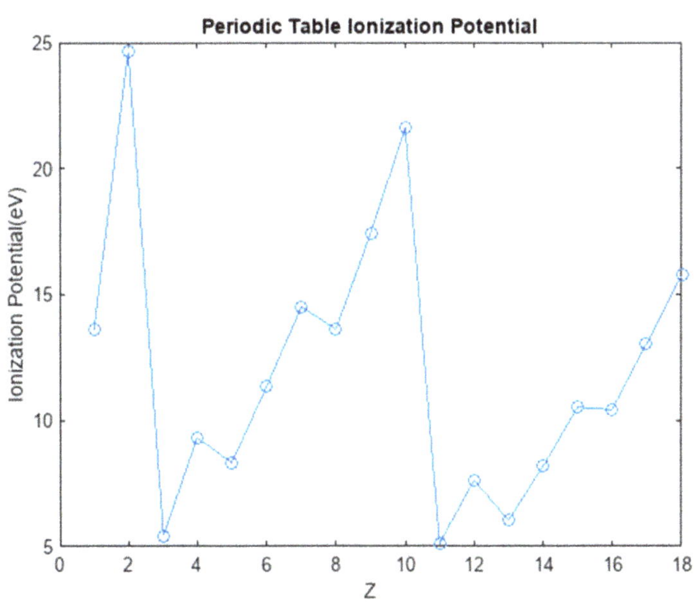

Figure 6.5: Ionization potential for atoms with Z from 1 to 18.

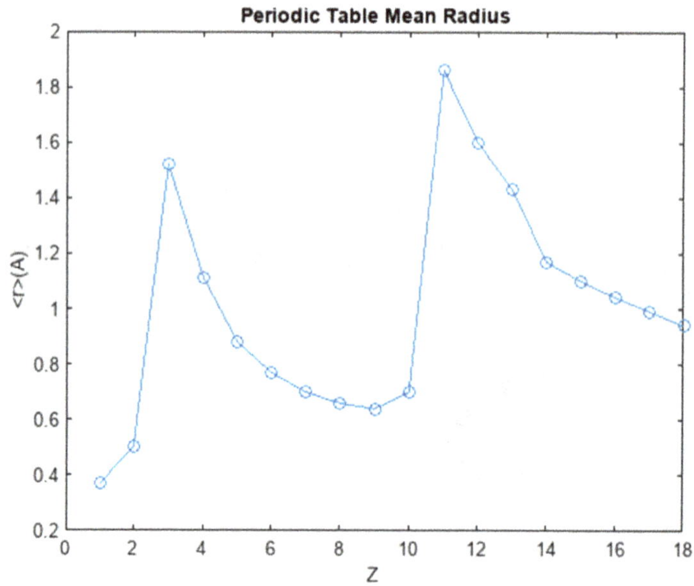

Figure 6.6: Mean radius of atoms with Z from 1 to 18.

He ($Z = 2$) is tightly bound (25 eV) but Lithium i($Z = 3$) is a metal. Then the 2s "shell" fills followed by the 2p "shell" ending in tightly bound Neon, a noble" gas. Then the 3s shell starts filling with lightly bound Sodium metal. Then the 3p "shell" starts to fill ending in another noble gas, Argon. Clearly the ionization potential data is a map of the shell structure of atoms Similarly as atoms become more complex they grow slowly in size, with maxima occurring with the lightly bound outer electrons of the metals and the tightly bound electrons of the bases and noble gases.

6.6 Energy Bands in Solids

Isolated atoms have sharp energy levels, for example the Hydrogen atom. In a solid, there are interactions with nearby atoms. This causes the allowed energy levels to expand into "bands". However, forbidden energies still exist. A simple model for the situation is a periodic potential due to the fixed ions, taken here to be at constant locations. The constant can be set by the user. The resulting energy levels for the shared electrons are displayed as is the lowest energy eigenfunction of the periodic potential.

This moves the discussion from the hydrogen atom, to complex atoms, and then to solids where the electrons have spread out allowable energies as they interact with the fixed ions in the crystal.

Band energy levels are crucially employed in the semiconductor industry. "Band Gap Engineering" is accomplished by adding small amounts of impurities to pure silicon or germanium crystals. These impurities have sharp energy levels within the band gap and they may be near the top (donor) or the bottom (acceptor) impurities.

The solution, called the Koenig-Penny model, is a free particle plane wave modulated by a periodic function $u(k)$. The width of the potential is b, while the length of the free space is a. The wave function and it's first derivative are required to be continuous at both boundaries, $x = 0, a$. Solving, there is a wave vector in region a and in region b, k_a, k_b. The solution is approximately:

$$0 \sim \cos(k(a+b)) - \cos(k_a a)\cosh(k_b b) - \left[\sin(\mathrm{ka})\sinh(k_b b)(k_b^2 - k_a^2)/2k_a k_b\right] \tag{6.3}$$

There is a forbidden zone where no solution exists. It occurs approximately when $k_a a = n\pi$. The constraint is not analytically tractable and is solved using "fminsearch" numerically. The discontinuity makes "fminsearch" a bit unstable, but the general trend is clear. The user chooses the ionic potential. The resulting band structure can be varied from $\sim$ free to larger band gaps.

```
% solve 1-d periodic potential, exact
% ions ~ constant potential
% energy in eV units
mec2 = 511000.0;  % eV - electron mass
hbarc = 2000.0;   % ev*A
% Electron of Energy < Vo, Bound in a Potential Vo Outside (0,a)
Where V = 0
% Periodic Potential = Vo for x  =(-b,0) and x = (a,a+b) etc.
% model for electron bound in  i-d ionic lattic
a =  4 ; %  Width of V = 0 (A)
b =  1 ; %  Width of Vo (A)
Vo = 5; % Ion Potential Vo (eV): ');
% approximate first forbidden energy (eV)
Ebad = (hbarc .^2 .*(pi ./a) .^2) ./(2.0 .*mec2)
```

```
Ebad = 2.4143
```

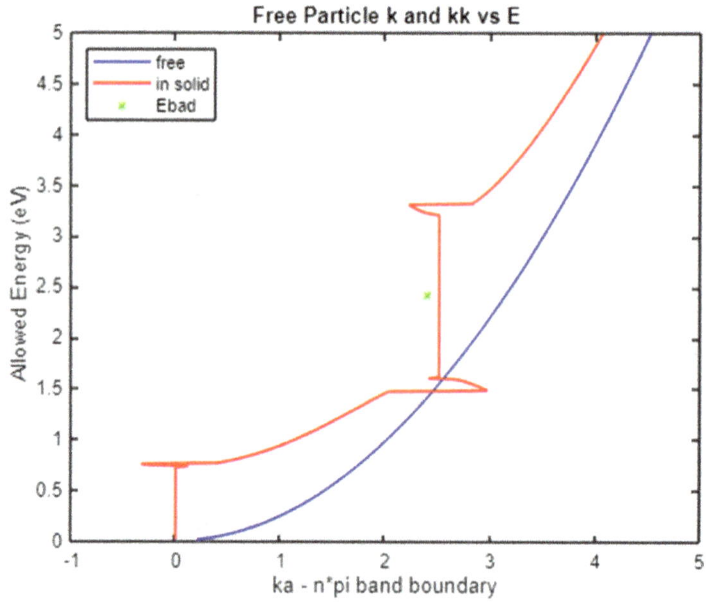

Figure 6.7: Energy of electrons bound in a periodic 1d lattice and for free electrons. Forbidden energy regions appear in the case of bound electrons.

The solutions are required to be periodic $u(a) = u(-b)$. The solutions are modulated plane waves. They are oscillatory for $x = (0, a)$ and exponentially damped for $x = (-b, 0)$

```
% now find eigenfunctions
% wave vector for V = 0, V = Vo
al = sqrt(2.0 .*mec2 .*EEE) ./hbarc;
be = sqrt(2.0 .*mec2 .*(Vo - EEE)) ./hbarc;
% find solved for k for this E- as close as possible
% 4 homogeneous eqs in 4 unknowns
% psi = uexp(jkx), u(x) = A*exp(j(al-k)x) + B*exp(-j(al+k)x)
%                         C*exp(be-jk)x)  + D*exp(-(be+jk)x)
% k solved by asking continuity of u and du/dx at x = 0 and a
% and asking periodicity, u(a) = u(-b)
```

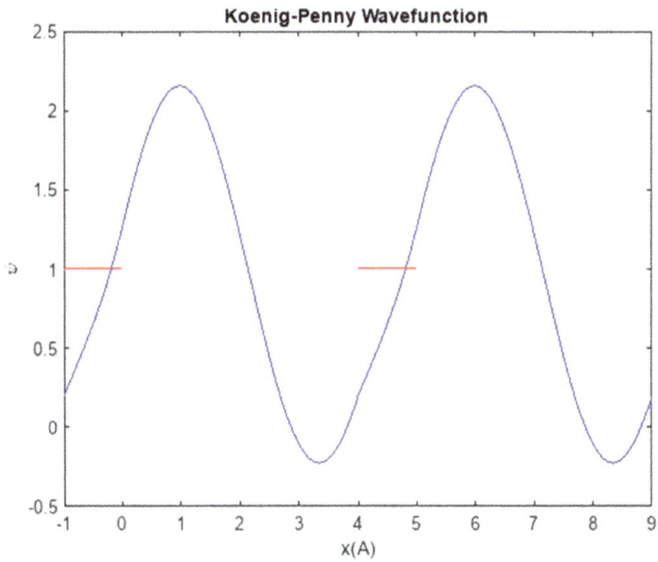

Figure 6.8: Wave function of electrons bound in a periodic 1d lattice. The location of the short range potential of the ionic nuclei are the red lines.

6.7 Simple Harmonic Oscillator

There is a closed form solution in quantum mechanics for the simple harmonic oscillator, a bound state like the hydrogen atom. In 1 dimension the solutions for the wave functions are Hermite polynomials and the quantized energies have a quantum number n, with associated energy $E_n = \hbar\omega(n + 1/2)$ where ω is the classical harmonic frequency. The quantum number n varies over the range of positive integers starting with 0. The lowest possible state has a non-zero energy, the "zero point energy". Indeed, at the lowest temperature a quantum oscillator has a non-zero energy.

The classical solution is plotted. The turning points are indicated, and the quantum solution is plotted for comparison. The quantum solution "leaks" outside the classical turning points. The user picks n. It is clear that for large n, the quantum solution begins to look like the classical one. This is called the "Correspondence Principle". For small values of n the difference are enormous, however.

The wave functions for small values of n can easily be written explicitly by asking for the Hermite polynomials. An electron is chosen as the bound mass. The solutions scale in position as n^2 times an exponential $e^{-y^2/2}$ where y is the dimensionless x position divided by the Compton wavelength of the electron, in this case with $\omega/c = 1$. The number of maxima for the wave functions is $n + 1$, while the mean position increases with n.

```
% Solutions to the Quantum SHO in 1-d
% SHO, k = 1, m = 1
% Choose the n value, En = hbar*w*(n+1/2)
% Shape of wave fuunction as a function of n
% approaches classical at very large n
n = 0; % Quantum # n = 0, 1, 2 , En in hbar*w units
e = (n + 0.5);     % hbar*w units
% classical turning points have v = 0, V = mw^2x^2/2
```

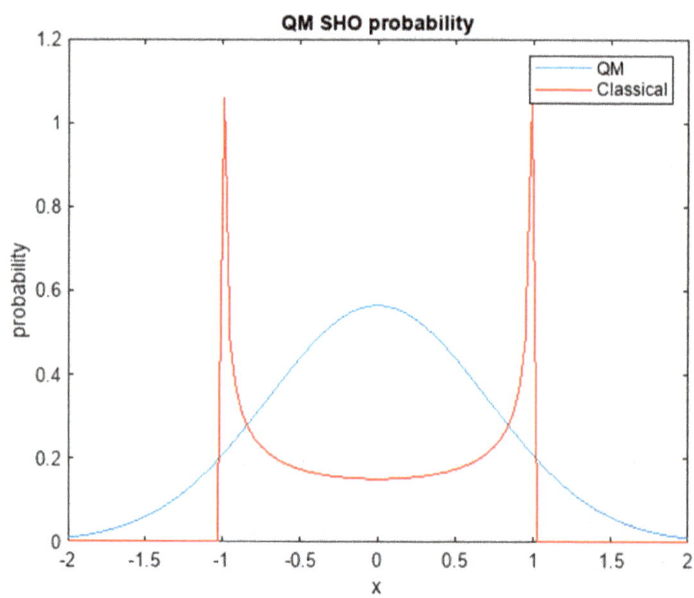

Figure 6.9: Wave function of the ground state of a quantum simple harmonic oscillator compared to the probability of observing the position of the classical oscillator.

```
%  wave function solutions for the Simple Harmonic Oscillator
% Initialize, electron bound
mec2 = 511000.0;  % eV - electron mass
hbarc = 2000.0;   % ev*A
% w/c = 1 in length units, y = x*sqrt(mec2*c/hbarc)*x
% Ground State Energy, w^2 = k/m, Eo = hbar*w/2, Energy Level
Spacing = 2*Eo
% Eo = hbar*w/2; ground state energy, w^2 = k/m (classical spring
result)
psi0 = (exp((-y .^2) ./2.0)) .^2; psi1 = psi0 .*4.0 .*y .^2; psi2 =
psi0 .*((4.0 .*y .*y - 2.0) .^2);
```

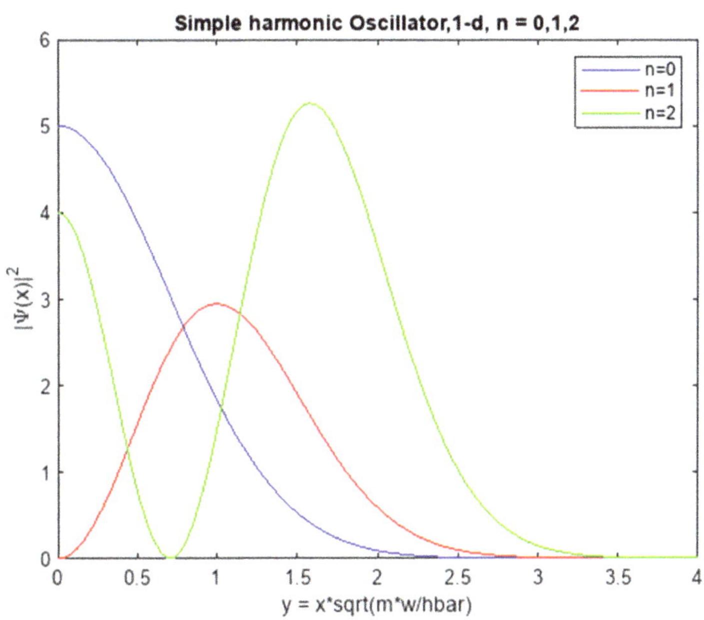

Figure 6.10: Wave function of a quantum simple harmonic oscillator for the first 3 principal quantum numbers.

6.8 Identical Particles

In classical physics there are independent particles, which can be considered to have joint probability distributions which are just the product of the individual distributions. In quantum mechanics the wave functions of 2 different particles can be treated independently as a product wave function. However, for identical particles, the spin/statistics properties are important. For 2 particles, the wave function describing the state must be symmetric (bosons) or antisymmetric (fermions) under interchange of the identical particle labels. The particles are, somehow, "entangled".

$$\psi_{a,b}(1,2) = [\psi_a(1)\psi_b(2) \mp \psi_a(2)\psi_b(1)]/\sqrt{2} \tag{6.4}$$

where a, b refer to the quantum state and 1, 2 refer to the particle labels. The following code looks at the simple case of 2 particles confined to a linear "box". The individual solutions are sin or cos functions which vanish at the boundary of the 1 dimensional box.

Particle 1 is in the ground state. The user chooses the quantum number for particle 2. Note that if $n = 1$ is chosen the fermion wave function vanishes due to the Fermi Exclusion Principle, 2 fermions cannot occupy the same quantum state (ignoring spin here). The wave function surfaces for the fermion and boson cases are both plotted. It is clear that quantum statistics has a fundamental influence of the wave function describing the 2 particle state. The fermions are position anti-correlated, while the bosons are positively correlated in position.

```
% look at two identical particles in QM
% n for 1-d particle confined to size a
% particle 1 has n = 1, cos(pi*x1/a)
% Particle 2 has cos(n*pi*x1/a), or sin, n <5
phi1 = cos(pi .*x1);
n = 3 ; % quantum number for particle 2 = 1,2,3, 4
% classically the solutions are product wave functions
% but for identical particles the overall wave function is symmetric
% bosons) or antisymmetric (frmions) under interchange
```

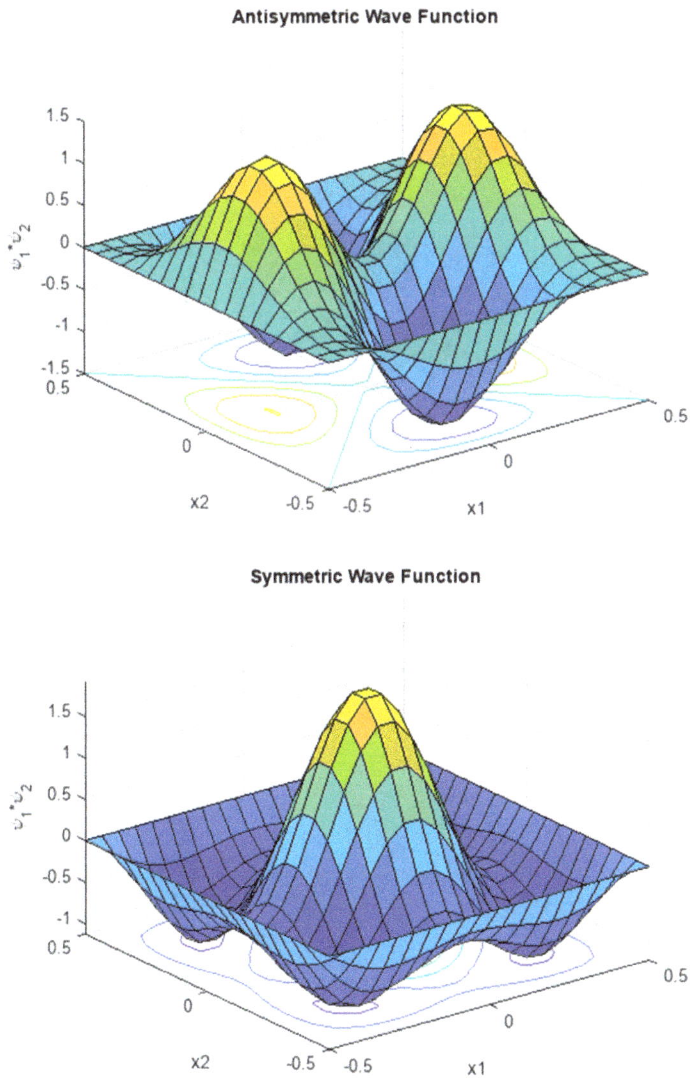

Figure 6.11: Wave function for a system of 2 identical particles. Top for fermions, bottom for bosons.

6.9 Scattering Off a Potential Well

In general in 1 dimension, a particle scatters off a potential well of depth Vo over a length a in x. The energy of the particle is fixed at 5 eV, but a and Vo are chosen by the user. The solutions are found by matching the wave function and it's derivative at the 2 boundaries, $x = 0$ and $x = a$. In general there is both a reflected wave and a transmitted wave, as well as left and right going waves inside the well. The matching is illustrated by plotting ψ for incident, reflected, transmitted, and inside the well. The ψ derivatives also match at the boundaries.

```
% Particle Scatters off a Potential Well of depth Vo, walls at (0,a)
% k = sqrt(2.0*me*(E))/hbarc; K = sqrt(2.0 .*me .*(E + Vo)) ./hbarc
% ao = pi ./K, lowest deep well state
% R = 0 at a Well Width = ao for this Depth
% R = 0 at Other Ramsauer Points x n(A) for this Depth
% algebraic solution - match wave function and derivative at
2 boundaries reflect
```

r1 = $-\dfrac{e^{-aki}\sin(Ka)(K^2-k^2)i}{-2Kk\cos(Ka)+\sin(Ka)(K^2+k^2)i}$

```
% transmit
```

t1 = $-\dfrac{2Kke^{-aki}}{-2Kk\cos(Ka)+\sin(Ka)(K^2+k^2)i}$

```
% in well, right
```

p1 = $-\dfrac{Kke^{-aki}e^{-Kai+aki}\left(\frac{k}{K}+1\right)}{-2Kk\cos(Ka)+\sin(Ka)(K^2+k^2)i}$

```
% in well, left
```

q1 = $\dfrac{Kke^{-aki}e^{Kai+aki}\left(\frac{k}{K}-1\right)}{-2Kk\cos(Ka)+\sin(Ka)(K^2+k^2)i}$

```
% exp inside well E < Vo
```

r2 = $-\dfrac{e^{-aki}\sinh(Ka)(K^2+k^2)}{\sinh(Ka)(K^2-k^2)-2Kk\cosh(Ka)i}$

t2 = $\dfrac{2Kke^{-aki}}{2Kk\cosh(Ka)+\sinh(Ka)(K^2-k^2)i}$

p2 = $-\dfrac{Kke^{-aki}e^{-Ka+aki}\left(\frac{k}{K}-1\right)}{2Kk\cosh(Ka)+\sinh(Ka)(K^2-k^2)i}$

q2 = $\dfrac{Kke^{-aki}e^{Ka+aki}\left(\frac{k}{K}+1\right)}{2Kk\cosh(Ka)+\sinh(Ka)(K^2-k^2)i}$

```
% Electron of Energy, E = 5 eV, Scatters off a Potential Well or
barrier
% Ramsauer if E =  -Vo +((hbar*k))^2/2*m, ka = n*pi
a = 5; % Well Full Width, a, in Angstroms
Vo = 3; % Well Depth Vo in eV, if <0 it is Barrier Height
% r is reflected amplitude
% p and q are + and - waves inside the well
% t is the transmitted wave
% probabilities for reflection/ trnasmission
RR = abs(r) .^2; TT = abs(t) .^2;
% find wave vectors , k,K
k = sqrt(2.0 .*me .*E) ./hbarc; K = sqrt(2.0 .*me .*(E + Vo))
./hbarc ;
% plot wave function modulus for incident, reflected transmitted
% note flux of reflected wave is negative - phase change on
reflection
```

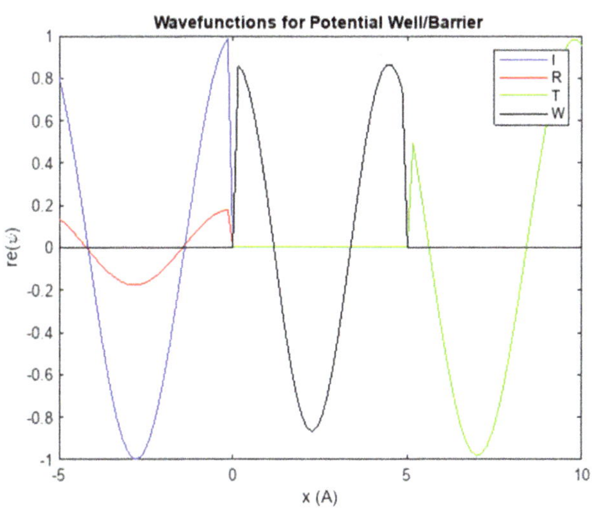

Figure 6.12: Wavefunctions for a potential well of finite width There is an incident wave, I. a wave reflected off the left well boundary, R, a wave transmitted through the well, T, and a wave inside the well, W.

Next the behavior of the reflection coefficient is plotted as a function of Vo and a for a fixed E value. It is well known in optics that one can manufacture an anti-reflection coating. Such coating are ubiquitous on camera lenses, for example. Indeed the plot below shows the points where R = 0 for the quantum square well. The cancellation of reflection is due to the destructive interference of the reflected wave and the left going wave which occurs when $a = n\pi/K$. The user can vary a and Vo and create such an antireflection condition.

```
% now find R,T for various Vo and a at fixed E see contour with
R = 0 antireflection
% and rapid falloff in tunnelling probability with a and Vo>Ee
```

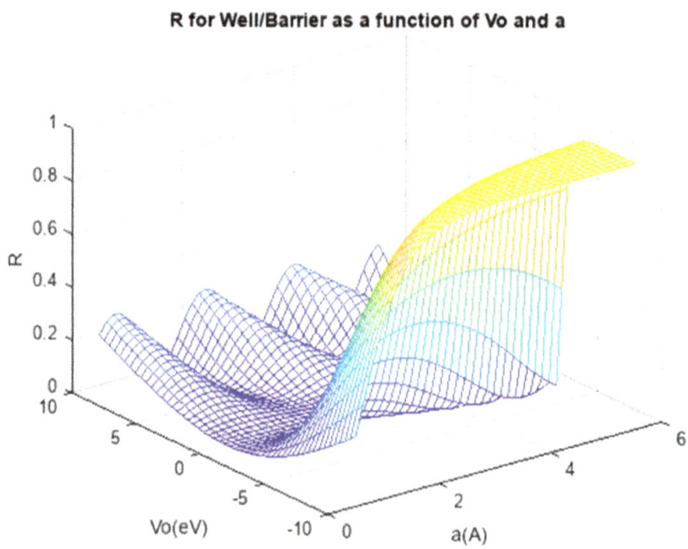

Figure 6.13: Reflected wave, R, as a function of the well depth (and height) and width. Points with R of both 1 and 0 are displayed.

6.10 Schrodinger Equation on a Grid

The time independent Schrodinger equation has a fixed energy solution, E. It then has the form of $-\hbar^2/2m\left(\frac{\mathrm{d}^2}{\mathrm{d}x^2}\psi\right) + V(x)\psi = E\psi$ in 1 dimension. This situation can be approached with the same techniques used in other equations. Specifically it can be solved on a grid. The solutions are called eigenfunctions and the energy for bound solutions has energy eigenvalues. The simplest solutions of problems with closed form solutions have been already explored, as in mechanics and electrodynamics, so now recourse to numerical methods is called for. The MATLAB utilities for matrices, "ones", "diag" are used to set up the matrix/grid for the problem and the utility "eig" then finds the eigenvalues and eigenvectors for the user. In this code the user gets to pick a form for the potential. However, any potential can be added as desired with a bit of editing. The potential is on the diagonal of the Hamiltonian matrix. Only the kinetic energy has off diagonal elements because of the necessity to approximate the second spatial derivative on the grid. As before the number of maxima of the wave function follows the principal quantum number.

```
% solve general time independent Schroedinger Eq
% for eigenvalues and eigenvectors - numerical solutions
neig = 3;  % number of solutions sought
% Pick A Potential Geometry;1 well, 2 wells, SHO, sinh V
const = ((hbarc .^2) .*N .^2) ./(L .^2 .*mec2);
% diagonal matrix element, add P^2/2m, with off-diagonal matrix
element
% setup Hamiltonian matrix, V + const
[phi,te] = eigs(H,neig,'SM'); % numerical eigenfunctions and
eigenvalues
% energy eigenvalues
```

```
e = 1x3
    0.3263    0.4128    1.0371
```

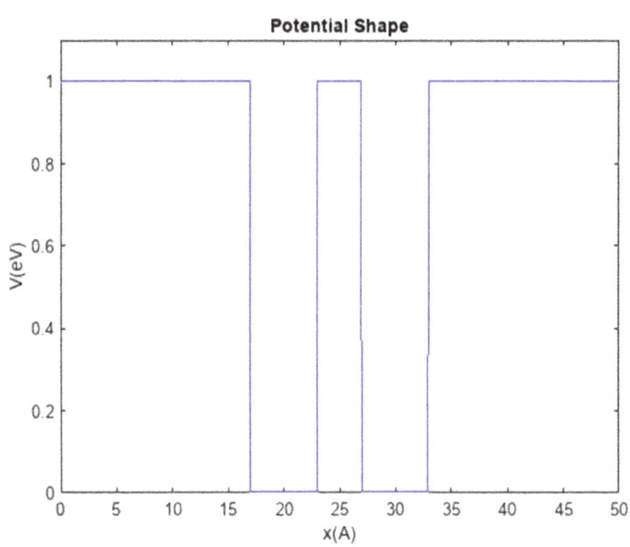

Figure 6.14: Potential with 2 distinct rectangular potential wells.

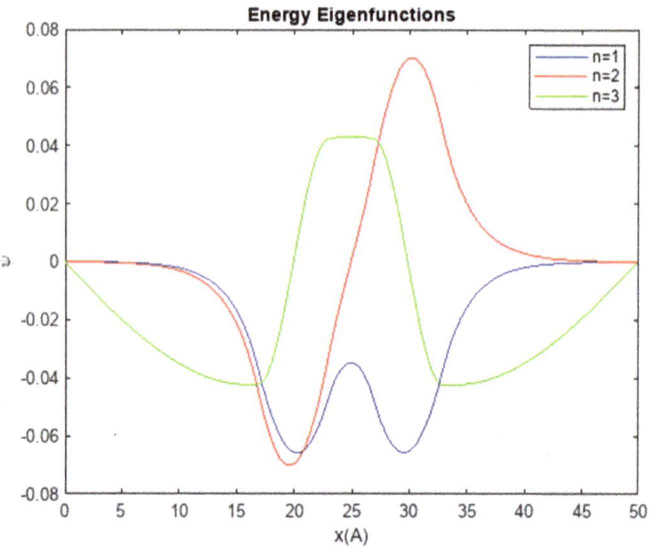

Figure 6.15: Wavefunctions for the case of a potential with 2 distinct rectangular potential wells.

6.11 Transmission, 2 Barriers

A solid crystal can be thought of as a series of barriers due to the fixed ions. The periodic potential was already explored in looking at allowed energy bands. Another approach is to approximately solve the problem for a series of thin potential barriers. There is a wide free space where the solutions are plane waves. There are also narrow regions where Vo is >E, and the solutions are exponentials. As usual the solutions are found by matching the wave function and the derivative at the boundary between regions. Assuming oscillatory motion with wave vector k, damped with wave vector K, a "propagation" matrix is used to connect right going waves, wave vector k, and left going waves, coefficients C and D at $-x$ to the coefficients of exponential wave vector K at $x > 0$ with coefficients A and B:

$$\begin{bmatrix} A \\ B \end{bmatrix} = \begin{bmatrix} ((1+\mathrm{iK})/k)e^{\mathrm{Ka+ika}} & ((1-\mathrm{iK})/k)e^{-\mathrm{Ka+ika}} \\ ((1-\mathrm{iK})/k)e^{\mathrm{Ka-ika}} & ((1+\mathrm{iK})/k)e^{-\mathrm{Ka-ika}} \end{bmatrix} \begin{bmatrix} C \\ D \end{bmatrix} \tag{6.5}$$

The 2 thin barriers have height Vo $= 7\,$eV. They are 1A wide (thin), separated by 5 A as default parameters. The user can choose both parameters, and study the resulting transmission coefficient as a function of the energy, E. There is quantum tunneling for E < Vo, where classically there should be full reflection, and there is reflection for E > Vo which is also not a classical effect. In addition there are specific energies where the reflection is 0, as was seen as the Ramsauer effect for the single well. As a simplification, the barriers are assumed to be quite thin; and Ka is set to 0 in the argument of the exponential.

```
% transmission through 2 barriers
% geometry
n = 2; % # of barrier
% Tunneling, E < Vo, Vo = 7 eV, fixed Vo
Vo = 7;
a = 6; % Width of V = 0 (A): ');
b = 1; % Width of Vo (~1) (A): ');
N = 2.0 .*n + 1 ; % potential boundaries to match
% scan energy % note e^Ka = 1 = e^-Ka approximation
```

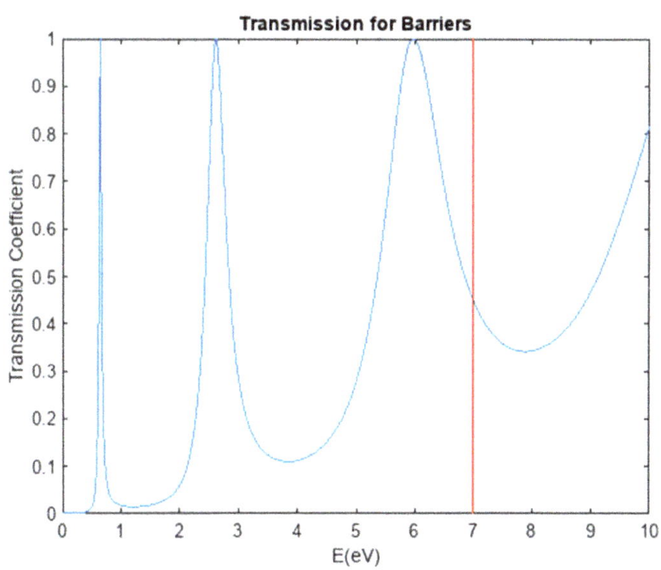

Figure 6.16: Transmission coefficient for a wave incident on 2 barriers with fixed height and width. Incident energies exist with $T = 1$.

6.12 Scattering in the Born Approximation

Physicists use the scattering of incident objects like electrons and targets like nuclei to study the forces acting between the projectiles and the target. The scattering angular distribution relates, in non-relativistic quantum mechanics, to the potential between them, $V(r)$. Spherical symmetry is assumed here. The scattering center is assumed not to recoil, and to be represented by a potential $V(r)$. For elastic scatters, the quantum mechanical Born approximation to the scattering angular distribution is:

$$d\sigma/d\Omega = (4m^2/\hbar^4 q^2)\left|\int_0^\infty \mathrm{rV}(r)\sin(\mathrm{qr})\mathrm{dr}\right|^2$$

$$q = 2k\sin(\theta/2), \quad E = (\hbar k)^2/2m, \quad d\Omega = 2\pi\sin\theta \mathrm{d}\theta \tag{6.6}$$

The projectile mass is m, momentum $\hbar k$, kinetic energy E, with momentum transfer q, while the solid angle is $d\Omega$ with a projectile scattering angle θ. Changing the variable to u = qr:

$$d\sigma/d\Omega = \left(\frac{1}{q^6}\right)\left|\int_0^\infty \mathrm{uV}\left(\frac{u}{q}\right)[\sin \mathrm{u}]\mathrm{du}\right|^2 \tag{6.7}$$

The basic power law is $1/q^6$ modulated by the expression which is squared and which contains the scattering potential. A simple example is a square well potential of strength Vo and range ro:

$$d\sigma/d\Omega = \left(\frac{\mathrm{Vo}^2}{q^6}\right)\left|\int_0^{\mathrm{qro}} u[\sin \mathrm{u}]\mathrm{du}\right|^2 \tag{6.8}$$

This example has an analytic solution which is derived below. In general the solutions are gamma functions. Another example is a power law potential in r, $\mathrm{V(r)} = 1/r^a$ which has analytic solutions for $1 < a < 3$, where $a = 1$ corresponds to electromagnetic or Rutherford scattering which is proportional to $(1/q)^4$

$$d\sigma/d\Omega - \left(\frac{1}{q^{6-2a}}\right)\left|\int_0^\infty u^{1-a}\sin \mathrm{u}\,\mathrm{du}\right|^2 \tag{6.9}$$

The following code covers these 2 examples and a "slider" is used so that the user can vary the basic parameter of the 2 examples. The first choice is the radial extent of the square well. A symbolic solution is displayed and the angular distribution for a chosen value of the radial extent of the square well is displayed. It is clear that optical like behavior is displayed in this case, with minima and maxima which depend on ro. In this case the shape of the cross section gives information on the radial size of the potential well, ro. A sharp boundary shows the oscillatory wave properties

```
% Look at NRQM elatic scatt in Born approx
% relationship of V to dsig/domega
% power law potential, square well, spherically symmetric,
exponential
% q = 2*k*sin(theta/2) T ~ k^2/2m
% dsig/domega ~ q^-2 [ int rV(r)sin(qr)dr]^2
% V ~ 1/r^a, a  = 1 Coulomb/Rutherford, dsig/domega ~ 1/q^4
% dsigma/domega ~ (1q^6)*|int(u*V(u/q)sin(u)du|2
% a < 3 - diverges a = 3. Define as 1`/q^6 times "modulation"
% start with square well, Vo = 1, ro,uo = q*ro, u = qr
```

ds_dom = $\dfrac{(\sin(q\ \mathrm{ro}\ ) - q\ \mathrm{ro}\cos(q\ \mathrm{ro}))^2}{q^6}$

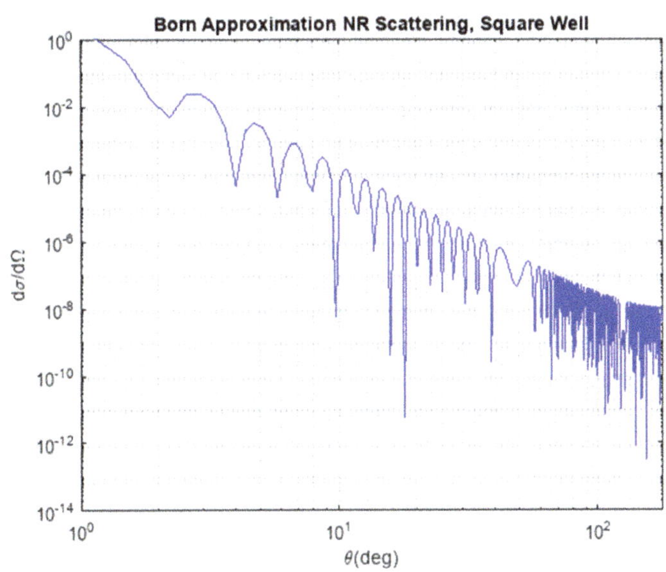

Figure 6.17: Differential scattering angle cross section for scattering off a square well in the Born approximation.

Notice that smaller ro well sizes lead to angular distributions peaking at larger angles. The scattering "measures" the size of the well. Note also that the structure in the "modulation" reflects the fact that there is a hard edge in V(r). Similar behavior occurs in classical wave scattering off extended objects.

In the case of V(r) falling as a power law in radius, the basic parameter is the exponent of r. Symbolic solutions are displayed which have factors with powers of q and gamma functions. The relationship between the power law exponent and the angular distribution is shown explicitly.

```
% now a power law falloff
% solution exists for al > 1 and al < 6 - the gamma function
% for a = 1 (Coulomb) ~ 1/q^4, Rutherford scattering
% for a = 3 ~ no q dependence ~ 1
aaa =1.06
```

$$\text{ds\_dom} = \frac{\cos\left(\frac{3\pi}{100}\right)^2 \Gamma\left(\frac{47}{50}\right)^2}{q^{97/25}}$$

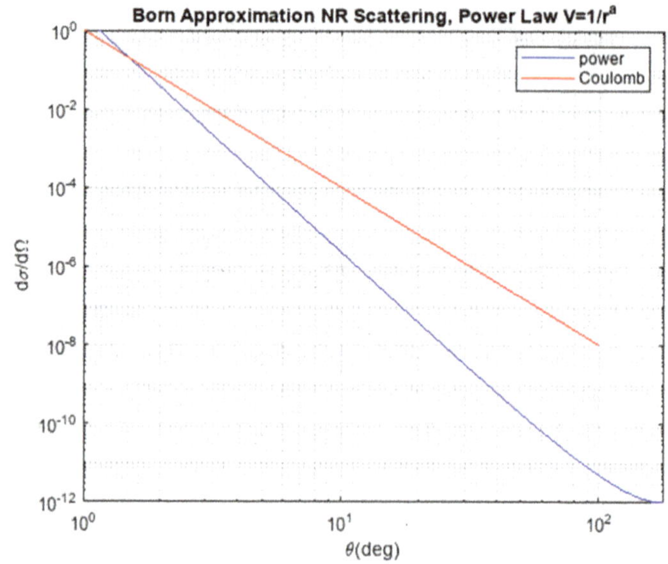

Figure 6.18: Differential scattering angle cross section for scattering off a power law point potential in the Born approximation. The blue curve power of q is 97/25. The red curve power is 4 which is numerically unstable.

In the case of a power law potential the angular distribution is smooth and monotonic. For a ~1 the q behavior is seen to be $\sim 1/q^4$ Coulomb scattering. For $a \sim 3$ the scattering becomes ~ isotropic. As before, larger values of a mean that the force is more localized at the origin and thus scatters more strongly at wide angles.

6.13 Wave Packets in 1-d, Scatter or Bound

Scattering

In quantum mechanics, matter has wave properties. For an energy E the momentum p and wave number k are $p = \hbar k = \sqrt{2m(E-V)}$ with an effective index of refraction of $n \sim \sqrt{1 - V/E}$. If a change in V occurs there is a reflection coefficient R and a transmission coefficient T where R + T = 1. If $E > V$ there are oscillatory solutions in the well while if $E < V$ the solutions are damped exponentials. This type of analysis is familiar from optics and carries over directly and easily into quantum mechanics. These analyses have previously been touched upon. Now, if scattering is to be studied a localized solution is needed in order to represent a "particle" and to follow the wave packet in time.

Scattering requires a representation of a "particle" which is consistent with the uncertainty principle. A "particle" in quantum mechanics is best represented as a wave packet with a range of wavelengths which allows the wave "packet" to be localized in space. The spread in wave number, dk, then is ~1/dx, so that more localized in space means more wave numbers are needed. The time spread due to the resulting energy spread is dt $\sim \hbar$/dE. An initial wave packet is constructed. It will spread in space with time even in the absence of interactions because of the spread of momentum values required by the Heisenberg uncertainty principle.

The following code uses specific units. Numerically E is in eV units, dt $\sim$ 0.66 in 10^{-15} s time units, and k in inverse Angstrom units is $k = 250/\beta$ where $\beta = v/c$. In these units E in eV is $3.9k^2$ for electrons with mass 0.511 MeV. These units are chosen so as to all be of magnitude ~1 for atomic electrons. Other units may be encountered elsewhere, for example nm which is, 1 nm = 10 Angstroms. In the scattering process the central electron energy is ~25 eV, with a spatial width of 1 A, with a corresponding spread in momenta/wave number, and with an initial maximum located at −10 A. The constant potential has a width of 5 A and a value Vo set by "slider" which can be positive ("barrier") or negative. Setting Vo to 0 allows one to observe the irreducible free particle wave packet spreading.

The Schrodinger equation in 1 dimension is solved for using the MATLAB utility "pdepe" for partial differential equations. This utility only covers 1 spatial and 1 temporal dimension.

```
% solve 1-d Schroedinger Eq using MATLAB PDE solver, Constant V,
well or
% barrier, scattering
% energy in eV units, length in A and time in 10^-15 sec
mec2 = 511000.0;  % eV - electron mass
hbarc = 2000.0; hbar = 0.666 ; % ev*A , eVto = 10^-15 sec
% Electron of Energy 5 eV, Scatters off a Potential Vo at (0,a)
% Initial Wave Packet, <x(0)>, dxo and k, x axis +- 20 A
% setup minimum uncertainty wave packet
dxo = 1.0;  % Wave Packet Spatial Spread dx(0) (A)
% evaluate expected spreading time - dx = dxo(1+(t/tau)^2)
% Expected Time for Packet Spreading in 10^-15 sec Units
% Wave Packet Initial Mean Location <x> (A), Grid of +- 20 A
bet = 0.01 ;%  Wave Packet Initial Velocity w.r.t. c, beta = v/c
% find wavenumber, energy - free particle
k = (mec2 .*bet) ./hbarc ; % k in 1/A-wave number, E=in eV
E = (hbarc .^2) .*k .*k ./(2.0 .*mec2)
```

```
E = 25.5500
```

```
sol = pdepe(m,@Sch_pde,@Sch_ic,@Sch_bc,x,t); % initial, boundary
conditions
```

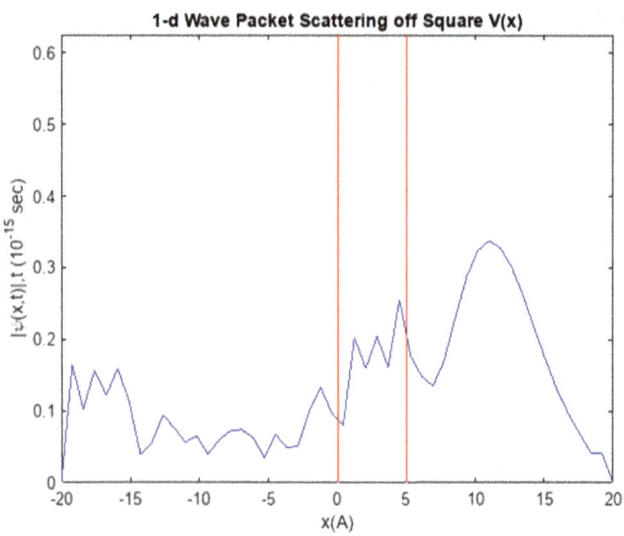

Figure 6.19: Last frame of a movie of a 1d wave packet scattering off a square well potential.

The potential can be attractive, Vo $<$ 0, or repulsive Vo $>$ 0. The height, Vo, can be $<E$ or $>E$. There is no really complete transmission nor reflection because the localized wave packet contains a range of energies. The reader is encouraged to run a variety of Vo values and to watch the behavior of the wave packet. If $E \gg$ Vo the transmission approaches the classical value of 1. However, even if Vo is <0, there is still some reflection. This is the same situation as in optics. Any change in the index of refraction results in both transmission and reflection.

"Bound" states with V constant well or SHO

The wave packet can also be placed inside a well of variable depth. Probability will "leak" out of the well even if the "particle" were thought to be confined by a potential $>E$. Approximate full confinement only occurs when the potential becomes very large. In the example below, $E = 25.5\,\text{eV}$ with a potential of $0\,\text{eV}$ inside the well and Vo outside the well with a full width 4 A. That situation is approximated by an harmonic oscillator potential here. The wave packet has a centroid at the origin with a spread of 1 A. The user picks the type, potential well or SHO and for the well also picks Vo

```
% solve 1-d Schroedinger Eq using MATLAB PDE solver, Constant V,
well or
% SHO, bound states
% Electron of Energy 5 eV, Bound in a Potential Vo > 5 eV Outside
(-a,a) Where   V = 0
% Initial Wave Packet, Centered at x = 0, grid +- 20 A in x
             % start in center of well
% find the constant potential for this problem Vo, or take SHO
% potential, V = x^2
% Pick Itype = 1, Vo or = 2 SHO
sol = pdepe(m,@Sch_pde2,@Sch_ic,@Sch_bc,x,t);
% eq defined, initial coonditions defined and boundary defined
```

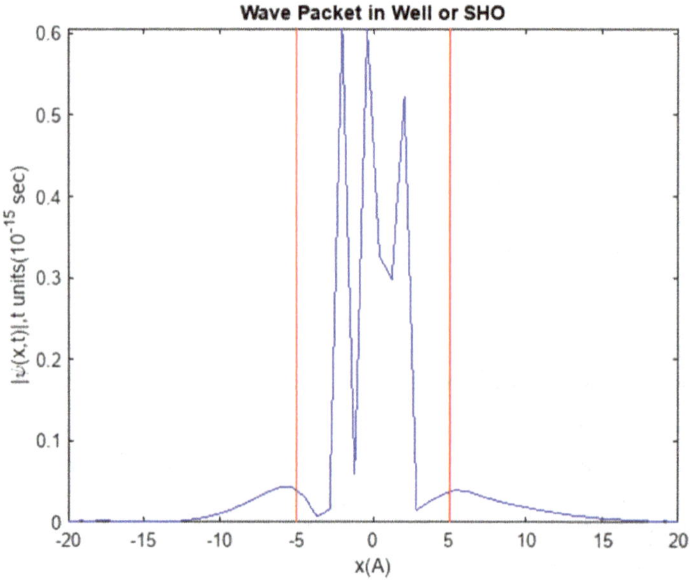

Figure 6.20: Last frame of a movie of a 1d wave packet placed inside a square potential well.

For a particle classically bound in a well $E <$ Vo, some probability is still lost. Only for the SHO potential, where the $V(x)$ diverges as x^2, is there containment of all the probability.

6.14 Wave Equation in 2-dimensions on a Grid

Partial differential equation solvers in 2 space and 1 time dimension can be easily written without using "packages". That approach may give more insight into issues of speed, convergence and stability. In any case the script that follows uses simple Gauss-Seidel with finite differences to numerically approximate the second order partial spatial derivatives in 2 dimensions and a look ahead numerical difference for the first order time derivative.

The Schrodinger time dependent wave equation is:

$$i\hbar\partial\psi/\partial t = E\psi = [-\hbar^2/2m(\nabla^2\psi) + V(x,y)\psi)] = E\psi \tag{6.10}$$

Treating the imaginary and real parts of the wave function separately, the time step n is approximated by a finite step with iteration n in time of Δt

$$\psi = \psi_R + i\psi_I$$

$$(\psi_R)^{n+1} = (\psi_R)^n + (\Delta t/\hbar)[-\hbar^2/2m(\nabla^2 + V(x,y)\psi_I)] \tag{6.11}$$

$$(\psi_I)^{n+1} = (\psi_I)^n - (\Delta t/\hbar)[-\hbar^2/2m(\nabla^2 + V(x,y)\psi_R)]$$

The spatial variations of the equation are approximated using a grid of points with x index j and y index k with grid spacing Δx and Δy respectively. This is the simplest estimate of the second order partial derivative in 2 spatial dimensions.

$$\begin{aligned}\nabla^2\psi &= [\psi(j+1,k) - 2\psi(j,k) + \psi(j-1,k)]/\Delta x^2 \\ &\quad + [\psi(j,k+1) - 2\psi(j,k) + \psi(j,k-1)]/\Delta y^2\end{aligned} \tag{6.12}$$

Scattering of 2d Wave Packets:

There are many options that the reader can explore. The first is s simple free wave packet. It spreads out in both dimensions as expected. A barrier, $V > E$ can be chosen with reflections observed. An attractive well is another possible choice.

Diffraction effects are seen when either a single slit or double slit located at $x = 0$ is chosen. The classical quantum 2 slit interference pattern is observable. Finally, both attractive and repulsive Coulomb scattering can be chosen. In all cases, the "impact parameter" can be varied by choosing a non-zero value for y_o. There are many choices possible and insights to be drawn from those choices. Some appreciable time spent exploring all the possibilities should be amply rewarding.

Bound States in 2d:

As in the 1 dimensional case, bound states can be approximated as a packet inside a potential. For a constant well, the packet will leak outside the boundary. For a SHO, the potential is truly confining.

```
% QUANTUM MECHANICS, Gauss-Seidel for spatial second order partial
derivatives
% Finite Difference Time Development Method, stability factor F
% x,y grid (0,1), (180 x 180), packet center (xo,yo), boundary
psi = 0
% The variable flagU (1,2,3,4,5,6,7,8,9) is used to change the
potential
% energy, or packet location
% 1 free propagation /2 Potential barrier /3 Potential Well
% /4 Single Slit /5 Double Slit /6 Coulomb potential (+Q)/7
Coulomb (-Q)
% /8 packet "bound" in a well /9 packet bound in a SHO potential
% Ian Cooper - initial coding
% movie frame time counter start
dt = 1.0e-5; % Time increment, stability ~ Fo ~ 0.16
%  setup [2D] GAUSSIAN wave packet - incident "particle"
x0 = 0.2 %  scattering, start at left side of box, xo = 0.2, bound
state xo = 0.5
```

x0 = 0.2000

```
y0 = 0.7 % impact parameter` box (0,1) x (0,1) b =0 at yo=0.5,
Change impact parameter b = yo  [ 0.2, 0.4 0.0.5]
```

y0 = 0.7000

```
% GAUSSIAN PULSE (WAVE PACKET) at t = 0
% Envelope
  psiE = A*exp(-(x-x0).^2/s).*exp(-(y-y0).^2/s);
% Plane wave propagation in +X direction
  psiP = exp(1i*k0*x);
% Wavefunction
  psi1 = psiE.*psiP;
% Extract Real and Imaginary parts
  R1 = real(psi1);  I1 = imag(psi1);
% Constant in S.E. analogue of Fo in temp flow - time stability,
f ~ 0.16
  f =  dt/(2*dx^2);
%  Note free packet spatial spreading in time
% 1 free / 2 barrier / 3 well / 4 single slit / 5 double slit /
6 Coulomb Q+/ 7 Coulomb Q-/8 packet "bound" in a well /9 packet
bound in a SHO potential
```

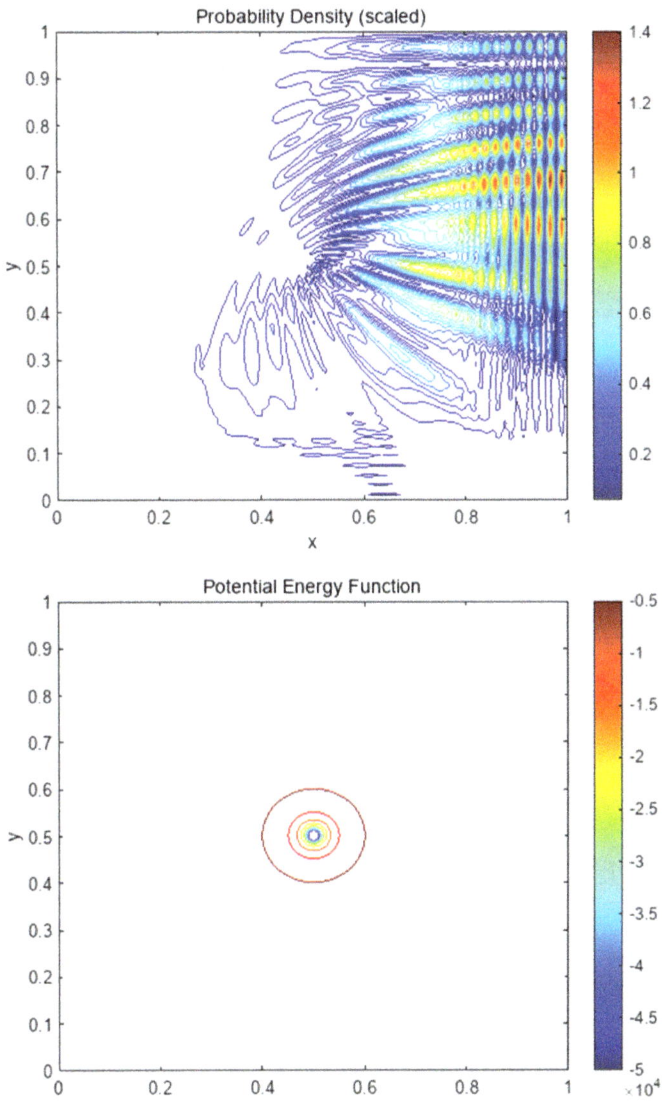

Figure 6.21: Top: Last frame of a movie of a 2d wave packet with an impact parameter scattering off an attractive Coulomb potential. This plot is the quantum version of Figure 1.7. Bottom: contours of the potential.

6.15 Decay Chain

A single decay follows an exponential law. In more complex situations found in radioactive dating there may be several sequential decays to be considered. A simple script is written to illustrate the situation with 3 sequential decays. The initial species, a, decays exponentially, which supplies species b, with parents for the b decays which are depopulated by the subsequent b decays, and so on. The equations are solved symbolically:

$$\begin{aligned} \mathrm{da/dt} &= -\mathrm{fa} \\ \mathrm{db/ct} &= \mathrm{fa} - \mathrm{gb} \\ \mathrm{dc/dt} &= \mathrm{gb} - \mathrm{hc} \end{aligned} \tag{6.13}$$

The following script allows the user to choose the b and c decay lifetimes. The intermediate b decay can be a check point on the c decays, as can be explored by changing both the b and c decays constants.

```
% Decay Chain - a -> b -> c with rates f, g and h
ode1 = diff(a) == -f*a; ode2 = diff(b) == f*a - g*b; ode3 = diff(c)
== g*b - h*c;
[odes] = [ode1; ode2; ode3];
cond1 = a(0) == 1; cond2 = b(0) == 0; cond3 = c(0) == 0;
conds = [cond1 ; cond2 ; cond3];
[asol(t), bsol(t), csol(t)] = dsolve(odes , conds);
asol(t)
```

ans = e^{-ft}

```
simplify(bsol(t))
```

ans = $\dfrac{fe^{-gt}}{f-g} - \dfrac{fe^{-ft}}{f-g}$

```
simplify(csol(t))
```

ans = $\dfrac{fge^{-ft}}{(f-g)(f-h)} - \dfrac{fge^{-gt}}{(f-g)(g-h)} + \dfrac{fge^{-ht}}{(f-h)(g-h)}$

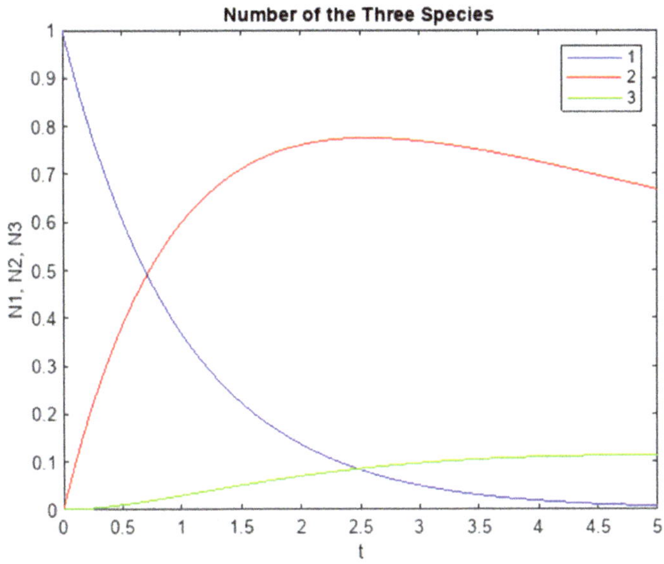

Figure 6.22: Number of 3 species in a sequential decay chain as a function of time. The parent is blue, the first daughter is red, and the second daughter is green.

Chapter 7

Special and General Relativity

"God abhors a naked singularity." Stephen Hawking

"In some sense, gravity does not exist; what moves the planets and the stars is the distortion of space and time." Michio Kaku

The rise of relativity theory occurred in the twentieth century as did quantum mechanics, as touched on in the previous Section. The first 5 Sections of this text looked at selected topics in classical physics; mathematical tools, mechanics, electromagnetism, gas and fluid flow, and optics and waves were all in place by the nineteenth century at the latest. The following Section looks at astrophysics and cosmology, both area of physics that have recently expanded greatly driven by the new tools and data that became available in the twentieth century (for example Hubble in 1922). This Section begins with Special Relativity (SR) followed by General Relativity (GR). Both topics were born early in the twentieth century.

Special relativity is special because it refers only to the relationships of physical quantities for observers in different free fall inertial frames ignoring any accelerated reference systems or residual gravitational effects. In such frames the basic postulate is that the speed of light is the same in all such frames. Velocities to not add and cannot be compounded to create a velocity $>c$.

As previously alluded, some quantities in SR are invariant, that is, they are independent of inertial frame. Proper time, ds is such a quantity, $ds^2 = (cdt)^2 - (dz)^2$, $ds = (cdt/\gamma)$, $z = vdt$, as is rest mass, m. Some quantities are SR vectors and transform between inertial frames by

Lorentz transformation. Examples are position and momentum, x_μ, p_μ, $x_\mu = (\overrightarrow{x}, \text{ct})$, $p_\mu = (\overrightarrow{p}, \varepsilon/c)$. The transverse components of the vectors do not change under Lorentz transformation. Clearly, the E and B fields do change, and they are components of a field tensor in electromagnetic theory. In GR there is a matter tensor with components ρc^2, $\overrightarrow{p}$ having the units of energy density. In addition the velocity is not a SR vector, but dx_μ/ds is. In general one needs to add velocities correctly in order to relate velocities and accelerations in 2 different SR frames. One special case occurs when a proper acceleration a_o is observed in a frame where the objects have velocity β, in that case $a = a_o/\gamma^3$. This section starts simply with the Lorentz transformation of time.

7.1 Time Dilation

The basic tenet of SR is that the velocity of light is the maximum velocity of any material object. If more energy is put into a system the energy will increase, but contrary to the expectation that $v = \sqrt{2\text{mT}}$, the velocity does not increase past $v = c$. One consequence of the postulate that the velocity of light is the same in all inertial frames is that time is not universal but depends on the frame you are in. A conceptual clock sending a light pulse vertically as a "tick" and receiving it as a reflected "tock" is followed below. For an observer at rest with respect to the clock the time interval is shorter than the time observed when the clock is in motion. The light just has farther to go. The geometric factor for the increase is $\gamma = 1/\sqrt{1-\beta^2}$, $\beta = v/c$. The user can vary β and observe the difference in clock ticks.

```
% Illustrate Time Dilation With a Gedanken Clock
% "Clock" is a Light Source and a Mirror
b = 0.9;% Velocity of the Clock w.r.t. c
gamma = 1.0 ./sqrt(1.0 - b .*b)
```

```
gamma = 2.2942
```

```
% the clock rest frame light bounces in y only
lig = [0 0];  % location of light and mirror in S
mir = [0 40];
c = 4;
```

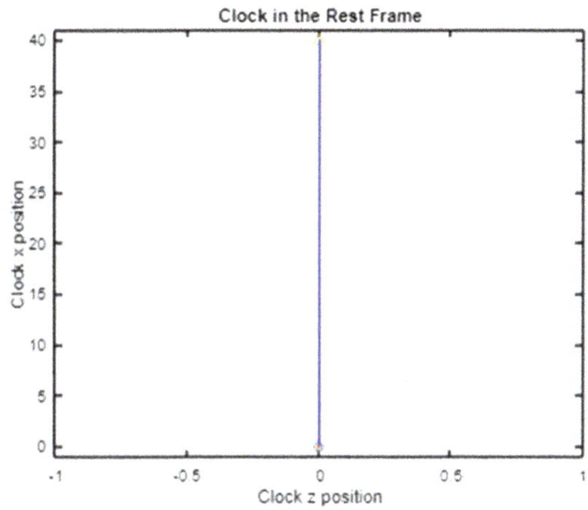

Figure 7.1: Schematic clock at rest with tick and tock the emission and reception of light.

```
jmax = floor(gamma .*imax)   % number of clock ticks in S' moving
frame
```

jmax = 45

```
% Clock Ticks in Rest Frame = 20, in Moving Frame =  jmax, longer
distance travelled
```

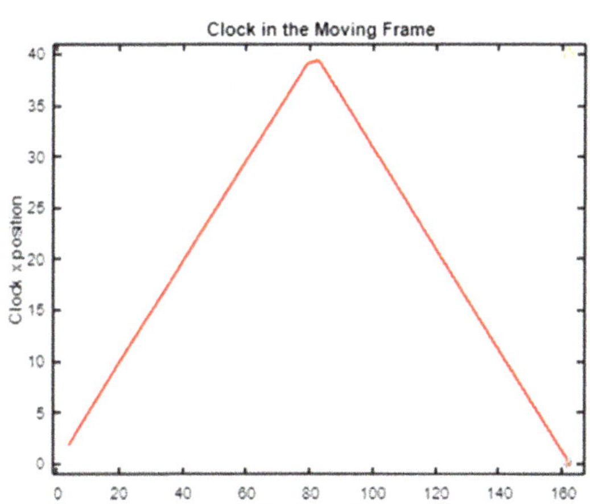

Figure 7.2: Schematic moving clock with tick and tock the emission and reception of light.

7.2 Twins and Rockets

Time dilation is a real phenomenon. Indeed, were it not for time dilation the muons produced by cosmic rays in the Earth's atmosphere would all have decayed before reaching the surface of the Earth, where they are observed to be copiously present. Another familiar consequence of SR is the twin "paradox". For amusement, assume a rocket could be built to provide a constant acceleration of 1 "g" — first class accommodations. Compare time for the stay-at-home twin (aka stationary mirror) and the traveling twin (moving mirror). This is a thought experiment to look at time dilation in a simple case, where the rocket has constant proper acceleration and the result is in terms of rocket (stationary or proper) time, s. In the rocket rest frame the acceleration is g. The stay-at-home observed acceleration in that special case is simple, $a = \mathrm{dv/dt} = (1-\beta^2)^{3/2}g$, where a v and t are all measured in the stay-at-home frame. Integrating over t, $\beta\gamma = \mathrm{gt}/c$. Solving for $\beta = (\mathrm{gt}/c)/\sqrt{1+(\mathrm{gt}/c)^2}$. In terms of the proper time, s, measured on the rocket, $\mathrm{ds} = \mathrm{dt}/\gamma = \mathrm{dt}/\sqrt{1+(\mathrm{gt}/c)^2}$ and integrating again, $\sinh(\mathrm{gs}/c) = (\mathrm{gt}/c)$. The controlling factor is the dimensionless variable $\Phi = \mathrm{gs}/c$. Expressions for β, γ, and z are derived similarly.

The velocity, time-at-home, and distance travelled for the stationary observer are:

$$\beta = \tanh(\Phi), \quad \gamma = \cosh(\Phi), \quad t = \sinh(\Phi)(c/g), \quad z = [\cosh(\Phi) - 1](c^2/g) \tag{7.1}$$

Note that it is only after about 2 years of constant 1 g acceleration that the SR and classical results begin to diverge at which time the traveler has gone about 2 lyr. However, at the end of 10 years the time ratio is ~1000. The user picks the ship time duration. Note that at constant acceleration, the distance travelled in light years is ~ the trip duration in years.

```
% rocket - Compute motion of a  relativistic rocket, constant
acceleration = g
g = 9.81;    % Gravitational acceleration (m/s^2)
c = 3.0e8 ; % c in m/sec
yr = 60.0 .*60.0 .*24.0 .*365.0;    % yr in sec
% express velocity, gamma, time and position of observer at rest -
ship
% time ts is proper time for ship
tsmax = 5; % Maximum Ship Time in Years
%
tss = (ts .*g .*yr) ./c ;
alf = (g .*yr) ./c ;  % time in yr, with g ==> alf ~ 1
```

```
bet = tanh(tss);
gam = cosh(tss);
tt = sinh(tss) ./alf;
zz = (cosh(tss)-1.0) ./alf; % distance in lyr
tt(imax) % Total Time Elapsed at Home (yr)
```

ans = 84.1168

```
zz(imax) % Total Distance Travelled in LY (Home)
```

ans = 83.1526

```
% get classical results too
```

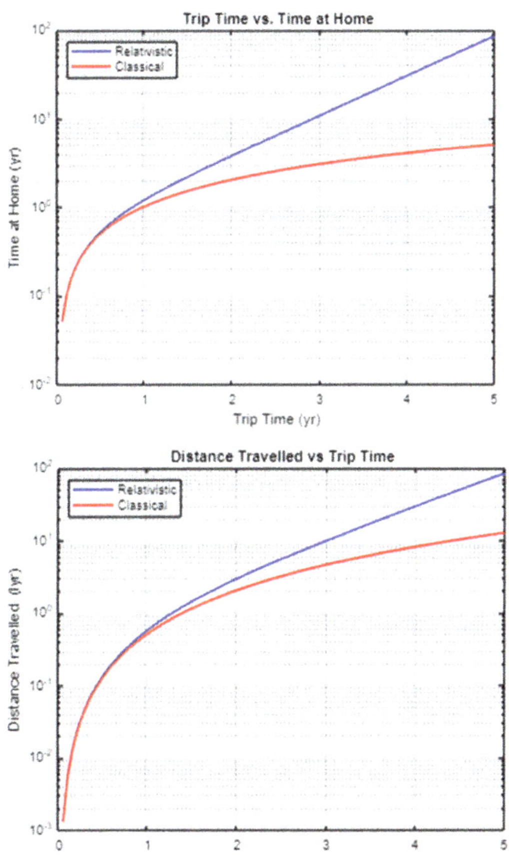

Figure 7.3: Time at home (top) and distance travelled (bottom) rocket with 1 g constant proper acceleration.

7.3 A Relativistic Rocket

First consider the rocket as an object of mass m which "decays" into an object of mass, m — dm and an ejected exhaust moving with velocity v_o with respect to the rocket. Conservation of both momentum and energy will be invoked in this process. Useful identities for the differentials are; $d\gamma = (\beta\gamma^3)d\beta$, $d(\beta\gamma) = \gamma^3 d\beta$. The modified rocket equation is: $m(d\beta/\mathrm{dm}) = (\beta_o/\gamma^2)$, $\beta = v/c$, which has the classical limit when the velocity is $\ll c$. This is the modification to the classical rocket where the change in β with mass is reduced by a factor $1/\gamma^2$ taking SR into account. Basically that factor takes into account that it is harder to increase velocity it you are already moving very rapidly, with a velocity which has a good fraction of c. This treatment is in terms of, not the rocket rest frame, but in the "stay-at-home" frame where the rocket has velocity β. The expression can be inverted to find the rocket velocity in terms of the initial and final mass of the rocket. Or one can express the mass in terms of the velocity.

The solution is found symbolically as $\beta(m)$ which depends on the exhaust velocity and the mass, so the payload fraction is important as expected. A specific example is set up, with a 10^6 kg initial mass, and a 40% payload ratio, although the payload fraction is under the control of the user. The exhaust velocity is also chosen by the user. However, it is clear from the $\beta(m)$ expression that the optimal exhaust velocity is c and photons are indicated as the propellant. The user is encouraged to play with the variable in order to see if relativistic rocket travel within the galaxy is at least theoretically feasible. The default settings keep the proper acceleration maximum $\sim g$. For photons the mass ejected is the E/c^2 value. The assumed rate is 10 gm/s of photons and the technology to accomplish that is left as an exercise for the reader.

```
% Relativistic Rocket Velocity, m*db/dm = bo*(1-b^2)
% symbolic solution
syms m b(m) mo bo mi
ode = diff(b(m)) == bo*(1-b^2)/m
```

$$\texttt{ode(m)} = \frac{\partial}{\partial m} b(m) = -\frac{\mathrm{bo}(b(m)^2 - 1)}{m}$$

```
init = b(mo) == 0;
bsol = dsolve(ode,init);
bb = simplify(bsol)
```

$$\text{bb} = \frac{m^{2\text{bo}} - mo^{2\text{bo}}}{m^{2\text{bo}} + mo^{2\text{bo}}}$$

```
% instead of b(m) one can invert to find m(b) with initial mass mi
% the result is: m(b)/mi = (1+b)^(1/2bo)/(1-b)^(1/2bo)
Mp = mfmo .*Moo; % payload mass
bo = 0.999; % Exhaust Velocity in Rocket rest Frame
b = abs( (1 - m .^alf) ./(1 + m .^alf)); % b(0) = 0
gg = 1.0 ./sqrt(1.0 - b .^2);
bf = b(1) % final beta
```

```
bf = 0.7237
```

```
gf = 1.0 ./sqrt(1.0 - bf .^2); % final gamma
```

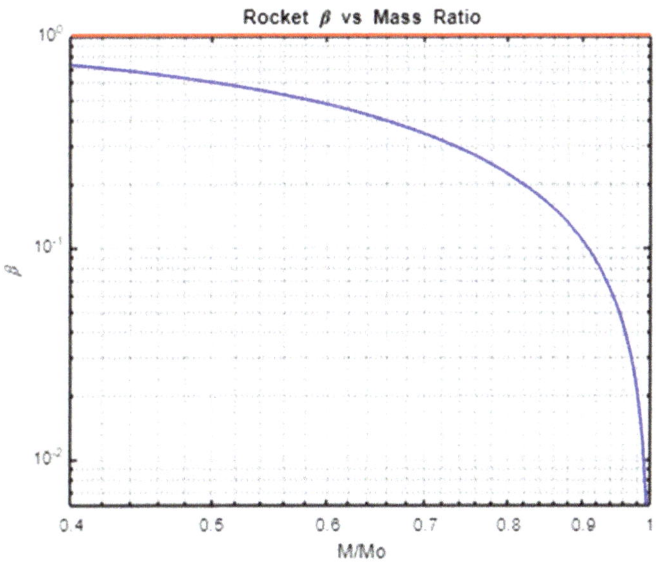

Figure 7.4: Rocket velocity as a function of the residual mass fraction (payload). The red line is the exhaust velocity.

```
% numerical integrate to get position, use "quad"
dbdt = (bo .*(1-b .*b) .*dmdt) ./M; % m is the fraction m/mo
aproper = (dbdt .*gg .^3 .*3e8) ./9.8 ;
aproper(1) % agmax, max proper acceleration in "g"
```

```
ans = 1.1078
```

```
yy(i) = quad(@sr_rock,0,t(i)); % integrate v to get position
```

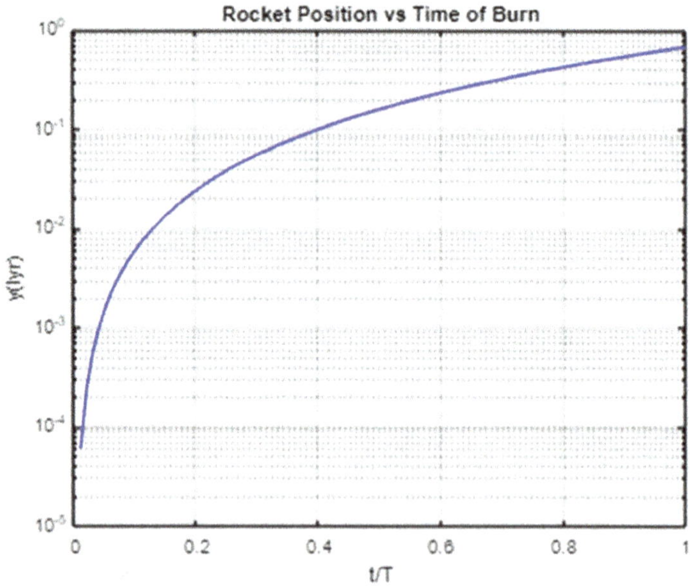

Figure 7.5: Rocket distance travelled as a function of fraction of the total burn time.

7.4 Reprise of Relativistic Units

Because of conventions, MKS and cgs units have both been used in this text. However, at present the majority of scientists use MKS units as the IS, International Scientific standard. It is easiest to use a consistent system of units. One also wants to use units appropriate to the scales of fundamental particles, and joules and kg are just too macroscopic. One common way forward is to always use eV units where 1 eV = energy gain in dropping through 1 Volt ($1.6 \times 10^{-19} J$). To continue, the conversion of particle masses from MKS units to eV is needed. The masses of the elementary particles e, π, K and p are approximately; 0.51 MeV, 0.14 GeV, 0.50 GeV and 0.94 GeV, 1 MeV = 10^6 eV, 1 GeV = 10^9 eV, 1 TeV = 10^{12} eV. These units will be consistently used in referring to fundamental particles. NR, UR, SR: refer to non-relativistic motion, UR to ultra relativistic motion, and SR to special relativity in general. There is only 1 other "velocity" variable which is needed to be chosen: v, β, or γ. The choice will be one of convenience.

Velocity variables:

$$\beta = \frac{v}{c} = \sqrt{1 - \frac{1}{\gamma^2}}, \quad \gamma = \frac{1}{\sqrt{1-\beta^2}} \tag{7.2}$$

Relationships between mass, m, energy, ε, and momentum, p, are:

$$\begin{gathered} p = \gamma\beta \mathrm{mc} = \gamma \mathrm{mv}, \quad \varepsilon = \sqrt{(\mathrm{mc}^2)^2 + (\mathrm{pc})^2} = \gamma \mathrm{mc}^2 \\ p = \frac{(\varepsilon\beta)}{c}, \quad \varepsilon = \frac{(\mathrm{pc})}{\beta} \end{gathered} \tag{7.3}$$

The velocity v is p/m for $cp \ll \mathrm{mc}^2$ or $\gamma \sim 1$ or $p \ll \mathrm{mc}$, that is NR motion.

$$v = \frac{\mathrm{dz}}{\mathrm{dt}} = \left(\frac{p}{\gamma m}\right) = \frac{\left(\frac{p}{m}\right)}{\sqrt{1 + \left(\frac{p}{\mathrm{mc}}\right)^2}} \sim p/m \tag{7.4}$$

The relationship between energy gain and momentum gain is not a constant. It is small for NR motion, and approaches 1 for UR motion. This kinematic result will be used later

$$\begin{gathered} \frac{\mathrm{d}\varepsilon}{c\mathrm{d}p} = \frac{pc}{\varepsilon} \\ \text{NR: } \varepsilon = \mathrm{mc}^2 + \frac{p^2}{2m}, \quad \text{UR: } \varepsilon = cp \\ \text{NR: } \frac{\mathrm{d}\varepsilon}{c\mathrm{d}p} = p/mc = \beta, \quad \text{UR: } \frac{\mathrm{d}\varepsilon}{c\mathrm{d}p} = 1 \end{gathered} \tag{7.5}$$

If masses are quoted in eV, or energy units, that is equivalent to setting $c = 1$. In the code calculations may be made in this way, but the quoted text will correct for this shortcut, even though it is widely used.

7.5 SR 2 Body Kinematics

The kinematics for 2 body scattering are similar to the case in NR mechanics except now the vector momentum and the scaler energy are separately conserved and E is not T but the total energy. The calculations are done in the center of mass, or better, center of momentum where the initial state momentum is zero (CM) frame where they are easiest and then the final state energy and momentum of the 2 outgoing particles is transformed back to the lab frame. The mass of the projectile is fixed at 1. First, consider a system of 2 particles. The mass, M, of the system is found using the expression ($c = 1$ here), $E^2 = p^2 + M^2$ so that $M^2 = (\varepsilon_1 + \varepsilon_2)^2 - (\overrightarrow{p_1} + \overrightarrow{p_2}) * (\overrightarrow{p_1} + \overrightarrow{p_2})$. The center of momentum system has velocity; $\overrightarrow{\beta_{\text{cm}}} = \overrightarrow{p}/\varepsilon = (\overrightarrow{p_1} + \overrightarrow{p_2})/(\varepsilon_1 + \varepsilon_2)$, $\gamma_{\text{cm}} = (\varepsilon_1 + \varepsilon_2)/M$. In the CM frame (* superscript here) the kinematics for a 2 body system are fairly simple for a decay $M \to 1 + 2$

$$\begin{aligned}\varepsilon_1^* &= [M^2 + (m_1^2 - m_2^2)]/2M \\ \varepsilon_2^* &= [M^2 - (m_1^2 - m_2^2)]/2M \\ M &= \varepsilon_1^* + \varepsilon_2^*, \quad p^* = p_1^* = p_2^*, \\ |p^*|^2 &= [M^2 + (m_1^2 - m_2^2)][M^2 - (m_1^2 - m_2^2)]/(2M)^2\end{aligned} \tag{7.6}$$

The exercise below is for 2 body scattering for a target at rest in the laboratory which scatters off a projectile. The user has a choice of the target mass at rest If it is 0 the projectile decays into particles 3 and 4. If it is non-zero there is a 2 body scattering and the target mass can be less than the projectile mass or more. The user can also pick the momentum of the projectile. Several distinct options can then be explored.

a) For a light target, mass <1, there is a maximum angle for particle 3
b) For a target mass of 1 a 90 degree recoil is possible, as in the classical case.
c) For a heavy target mass, mass >1, the projectile can be thrown backward.
d) For a decay the projectile with mass 2.5 decays into 2 mass 1 particles. There is a maximum transverse momentum for particle 3, and pz3 is always >0. There is a point where $\theta_3 = \theta_4$.

The user is encouraged to make several distinct choices for $m(2)$ — the target mass, and $p(1)$ the projectile momentum. By the way, the relativistic rocket problem done previously can be thought of as a strictly linear decay of a particle of mass m decaying into a rocket of mass m — dm and an ejected fuel of mass dm and velocity β_o in the rocket rest frame.

A simplified version of the kinematics, to serve as an example, is to consider the special case of a decay of a mass M moving in the lab with momentum p, γ, β into 2 photons. The decay is in a plane, taken to be (x, z). Analysis is easiest in a frame where M is at rest, the center of momentum frame. In that frame, labelled with a *, energy conservation requires that both photons have $\varepsilon_* = M/2$ and are back to back with angle θ^* with respect to the incoming direction of M. Then transforming the momentum from the M rest frame back to the lab; $p_{x1} = p^* \sin\theta^* = p_{x2}$, $p_{z1} = \gamma(\varepsilon_* \cos\theta^* + \beta\varepsilon_*)$, $p_{z2} = \gamma(-\varepsilon_* \cos\theta^* + \beta\varepsilon_*)$. The lab angles of the photons are $\tan\theta_1 = p_{x1}/p_{z1}$, $\tan\theta_2 = p_{x2}/p_{z2}$. A special case is when the center of mass angle is, $\theta^* = 90°$. In that case the lab angles are equal and opposite, The sum of the angles is the lab opening angle, $\theta_1 + \theta_2 = 2\tan^{-1}(1/\gamma\beta) = 2\tan^{-1}(\mathrm{Mc}^2/\mathrm{pc})$. More complex cases are covered in the code below.

```
% does 2 body SR kinematics, elastic 2 --> 2, 1+2 --> 3+4 or
1 decay, 1 -> 2 + 3 -  full SR
m(1) = 1; % incident
m(2) =0.95;  % target mass at rest, if m(2) = 0 decay, 1 --> 3 + 4
if m(2) > 0
    m(3) = m(1); % elastic scattering
    m(4) = m(2); % in and out, 1+2->3+4
end
%
if m(2) == 0
    m(1) = 2.5; m(3) = 1; m(4) = 1; % decay
end
p(1) = 3; % incoming pz of p1
% does kinematics as a function of cm cos(theta) from ctmin to ctmax
% masses for 1 + 2 --> 3 + 4 , p is lab beam momentum of 1, p2 = 0,
pc
% if decay, then m(2) = 0.0,  m(1) --> m(3) + m(4)
% find cm quantities, beam energy, cm energy, beta gamma of cm,
% momentum in the cm and outgoing energy of 3, 4 in the cm
e = sqrt(p .^2 + m(1) .^2); % all in c = 1 units, e = mc^2,
p = g*b*c*m
ss = m(1) .^2 + m(2) .^2 + 2.0 .*m(2) .*e ; % cm energy^2
s = sqrt(ss) % CM energy, beta and gamma
```

```
s = 2.8126
```

```
b = p ./( e + m(2)) ; % target at rest
g = (e + m(2)) ./s;
ps =(ss - (m(3) + m(4)) .^2) .*(ss - (m(3) - m(4)) .^2); % CM
momentum
```

```
ps = sqrt(ps) ./(2.0 .*s); % in cm p1 = p2 = ps
es3 = sqrt(ps .^2 + m(3) .^2); % CM energies of 3 and 4
es4 = sqrt(ps .^2 + m(4) .^2);
% CM decay/scatt angle
cts = linspace(ctmin,ctmax,400);
sts = sin(acos(cts));
% find lab energies and angles of 3 and 4
e3 = g .*(es3 + b .*ps .*cts);
e4 = g .*(es4 - b .*ps .*cts);
pt = ps .*sts; % pT
pp3 = g .*(ps .*cts + b .*es3); % pz
pp4 = g .*(-ps .*cts + b .*es4);
th3 = atan2(pt ,pp3); th3 = (th3 .*360) ./(2 .*pi); % theta -
rad -> deg
th4 = atan2(-pt ,pp4); th4 = (th4 .*360) ./(2 .*pi);
p3 = sqrt(pt .^2 + pp3 .^2); % p
p4 = sqrt(pt .^2 + pp4 .^2);
```

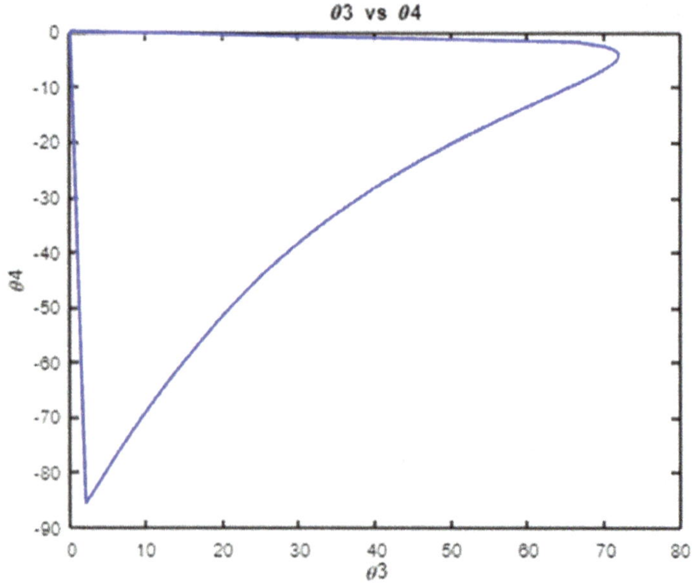

Figure 7.6: SR kinematics for $1 + 2 \to 3 + 4$ elastic scattering, $3 = 1$, $4 = 2$.

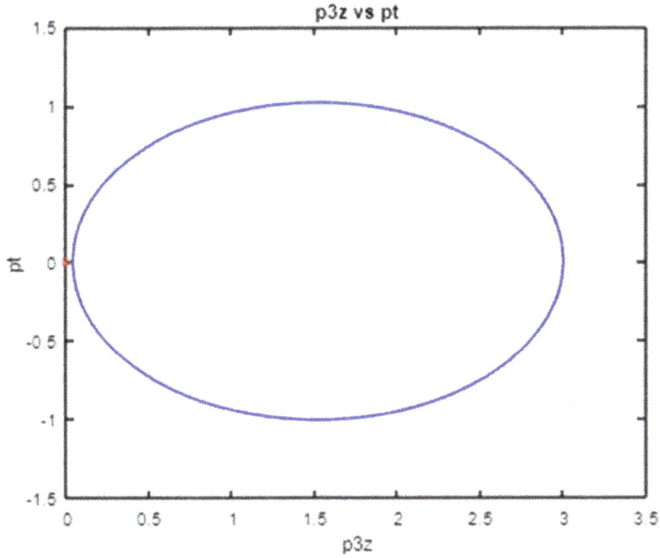

Figure 7.7: SR kinematics for $1 + 2 \rightarrow 3 + 4$ elastic scattering, $3 = 1$, $4 = 2$, $p1 = 3$.

7.6 Compton Scattering

Compton scattering showed that a photon scattering off a charged electron changed it's wavelength. The NR description, with no discernable wavelength shift, has already been discussed as Thomson scattering If it is accepted that a photon is a particle carrying energy and momentum, then it will lose energy in the scattering and since $E = \hbar\omega$ the frequency will decrease, or the wavelength will increase. The electron Compton wavelength, off which the photon scatters is $\lambda_c = h/m_e c$. The scattering can be treated purely kinematicly as a 2 body scattering process, using the conservation of vector momentum and energy. The photon can always backscatter, sending the electron forward. A forward going photon sends the electron at a 90 degree recoil. The user can vary the photon energy over a wide range in order to see the locus of points of the outgoing photon.

```
% Compton angular distribution as polar plot
% evaluated in target rest frame
% roughly isotropic in cm (1+ct^2) - NR
% wo = omega is the incident photon energy in MeV
% electron mass in MeV
me = 0.511 ;
```

```
% kinematics is 2-body elastic,relation of incident wo to scattered
photon energy w
% scattering angle of photons ct = cos(theta)
% x is ratio of photon energy to e mass, y is ratio of out/in
photon energy
slide =-0.4;
wo = 10 .^(slide)
```

```
wo = 0.3981
```

```
y = wo ./me ;
% backscattered photon energy
sb = wo ./(1+2.0 .*y)
```

```
sb = 0.1556
```

```
s = 1.0 ./(1.0 + y .*(1-ct)); % = w/wo
dsig = s .^2 .*(s + 1 ./s - (1-ct .*ct));
% dsig has a factor (alpha /m)^2 /2 not put in explicitly
% NR limit is dsig/domega --> (alpha/m)^2/2 *(1 + ct^2)
% NR sigma is Thompson result, 8*pi/3 *(alpha/m)^2
% polar plot - MATLAB utility
polar(t,dsig,'-b')
```

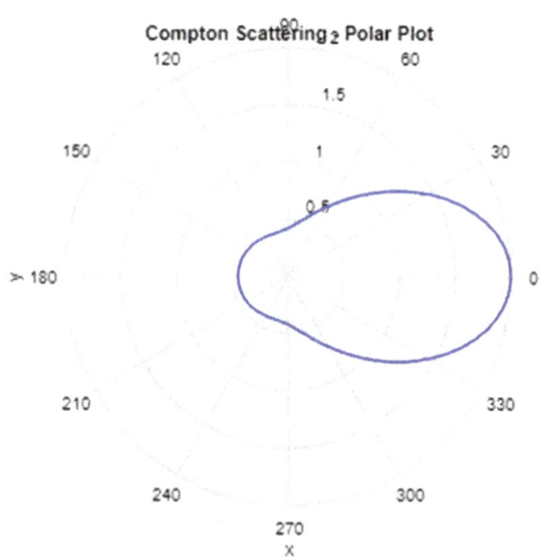

Figure 7.8: SR kinematics for photon scattering off an electron. The photon can backscatter, but typically goes forward.

```
% e recoil angle
phie = atan(1 ./((1 + y) .*tan(t ./2)));
phie = phie .*360 ./(2 .*pi);
```

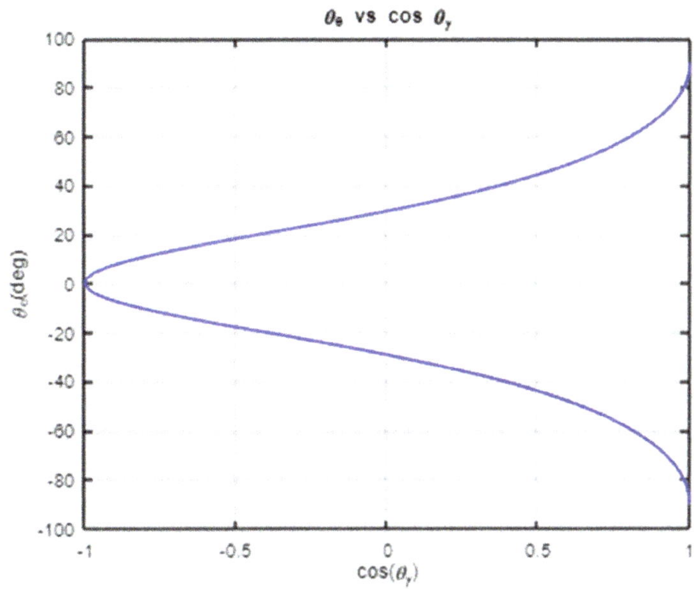

Figure 7.9: SR kinematics for a photon scattering off an electron. The electron recoil angle as a function of the cos of the photon scattered angle. Backscattered photons send the electron forward.

7.7 SR Electric Field

The beam in an accelerator is an ensemble in space and time of moving charged particles, a current. The electric field of this current is then detected by instruments used to measure and monitor the properties of the beam. Beam Position Monitors (BPM) are used to measure the effects of errors in the accelerator elements — quadrupole alignment, B field errors and quadrupole gradient errors. The BPM discussed in this text use the induced image charge of the beam on some conductor, assumed here to be near the beam vacuum pipe radius. First we need to know what the field of a moving charged particle looks like in SR. It does not behave like position or momentum do in SR.

The maximum NR electric force is $\sim(\mathrm{qe})E_o = (\mathrm{qe})^2/(4\pi\varepsilon_o b^2)$ where b is the impact parameter. The field acts over a time $\sim b/v$, so that the momentum impulse seen by the observer at $(x, z) = (b, 0)$ is $\Delta p_x = (\mathrm{qe})E_o\Delta t$. The effect of SR is to increase the field by a factor γ, but the relevant time shrinks by a factor γ, leaving the momentum impulse approximately constant. In the UR regime, the $E_x(t)$ pulse is almost a "Dirac delta function" acting strongly but only over a very short time (impulse). Moreover, the momentum transfer is quite constant over a wide range of velocity. All moving charges act like "minimum ionizing particles", "mips", because of this. The electric field at a present position (x, z) of the charge with velocity β has an electric field at the observation point $(b, 0)$

$$Q = (\mathrm{qe}/4\pi\varepsilon_o) \cdot E_x = Q\gamma\frac{b}{[b^2 + (\gamma\beta\mathrm{ct})^2]^{3/2}}, \quad E_z = -Q\frac{(\gamma\beta\mathrm{ct})b}{[b^2 + (\gamma\beta\mathrm{ct})^2]^{3/2}} \tag{7.7}$$

The Live script that follows makes a "movie" of $E_x(t)$ and $E_z(t)$ for values of β chosen by the user employing a "Numeric Slider". The pulse shapes at values of β near 1 are of most interest for accelerators and other SR applications. The symbolic fields E_x and E_z are shown first. The time integral of E_z is 0. The moving particle in the movie is a blue dot, while the field at the observer is a red line. Note that the transverse electric field is not the same in all reference frames, an invariant. In the past it has been assumed that transverse position and momentum are the same in all inertial frames, since they are components of a 4 dimensional vector. However, the E and B fields are not vectors, but components of a Field tensor, and the transverse electric field is very frame dependent. In the next Section the gravitational stress tensor will be discussed, with elements containing the mass density and the pressure.

```
% Electric Field of Moving Charge
Ex = gamma/(1+t^2)^(3/2);
Ez = -t/(1+t^2)^(3/2);
bet =0.98 ;
% tprime = t/to = t/v/b
tp = linspace( -5,5,200);
g = 1.0 ./sqrt(1.0 - bet .^2);
fact = 1.0 ./(1.0 + (g .*tp) .^2) .^1.5;
Ex = g .*fact;
Ez = - g .*tp .*fact;
```

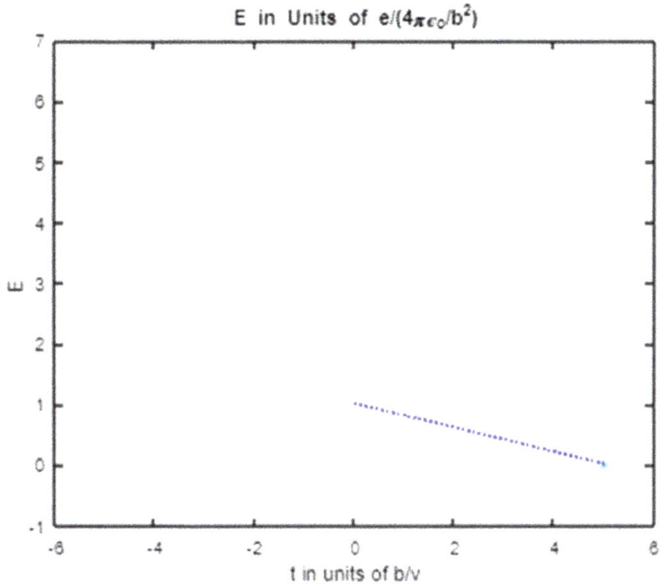

Figure 7.10: Last frame of the movie of a charge moving along the z axis past a point at $(0, 1)$.

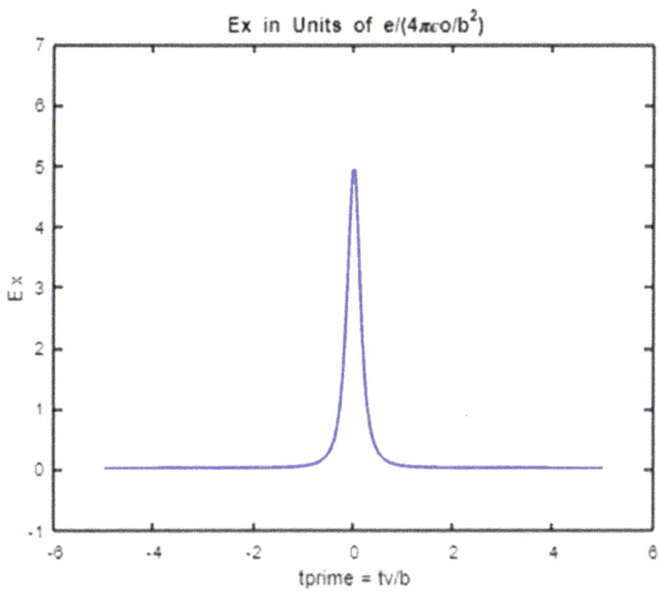

Figure 7.11: Transverse electric field at $(0, 1)$ as a function of scaled time.

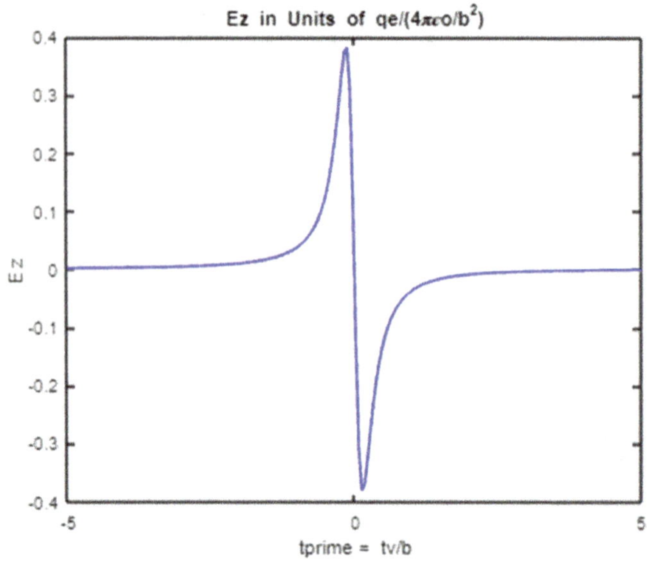

Figure 7.12: Longitudinal electric field at $(0, 1)$ as a function of scaled time.

7.8 Synchrotron Radiation

Starting with NR radiation by an accelerated charge, it is found to have an angular distribution in lowest multipole order which is dipole, with emission preferentially perpendicular to the acceleration (Larmor). Specifically, in a magnetic field, the Lorentz force equation implies a rotation of the charged particle which has the instantaneous circular frequency, $\omega_o = \mathrm{qeB}/(\gamma m) = \beta c/\rho \sim c/\rho$. The acceleration is, $a = c^2/\rho$ and the period is $\tau = 2\pi\rho/c$. In the NR case the radiation of a charge qe depends on the square of the acceleration a. and the fine structure constant α since there is one vertex where the electron emits a photon. Typically a process with n vertices has a cross section $\sigma \sim \alpha^n$, The time averaged radiated power, $\langle P\rangle$, is given by the Larmor expression for NR motion:

$$
\begin{aligned}
d\langle P\rangle/d\Omega &= (\alpha\hbar c/4\pi c^3)(\mathrm{qa})^2 \sin^2\theta \\
\langle P\rangle &= (\mathrm{qea})^2/(6\pi\varepsilon_o c^3) = (\mathrm{qa})^2(2/3)(\alpha\hbar c)/c^3
\end{aligned}
\tag{7.8}
$$

Electron beams can be extremely relativistic due to the low mass, 0.511 MeV, of the electron. A 50 GeV electron then has a γ factor of ~100,000 which certainly qualifies as UR. Indeed, it is the copious radiated power and the associated cost of supplying it, which currently limits the energy of electron circular accelerators.

First, consider a photon in 2 reference frames with uniform relative motion defined by β and γ. The relationship between photon angles for an isotropic distribution in one frame, the * frame, transformed to a distribution in a frame moving with velocity βc uses the SR transformation properties of massless photons for momentum p_z and energy $p(c = 1$ here), $p = \gamma p^*(1 + \beta\cos(\theta^*))$ and $\mathrm{pcos}(\theta) = \gamma p^*(\cos(\theta^*) + \beta)$. A code snippet gives the result below. A solid angle $d\Omega$, is $d\Omega = d\phi d(\cos\theta) = 2\pi d(\cos\theta)$ with integral $= 4\pi$. Typical photon angles are thrown forward into a cone limited by $\theta \sim 1/\gamma$, as mentioned previously.

Specializing to circular motion with velocity β, along z, perpendicular to $d\beta/\mathrm{dt}$, along x instantaneously, the time averaged radiated energy, $\langle U\rangle$, per solid angle and frequency is, in the small angle approximation

$$
\begin{aligned}
&1 - \beta\cos\theta \sim (\theta^2 + 1/\gamma^2)/2 \\
&d^2\langle U\rangle/d\Omega d(\hbar\omega) \\
&\quad = (\alpha/4\pi^2)(\omega/\omega_o)^2(2/3\gamma^2)^2(1+y^2)^2[K_{2/3}^2(s) + (y^2/(1+y^2))K_{1/3}^2(s)] \\
&\text{where } y = \theta\gamma, \quad s = (\omega/(3\omega_o\gamma^3)(1+y^2)^{3/2}) = (\omega/3\omega_c)(1+y^2)^{13/2}, \\
&d(\cos\theta) = d(\cos\theta^*)/\gamma^2(1-\beta\cos\theta^*)^2
\end{aligned}
\tag{7.9}
$$

```
% snippet for searchlight, transform from * frame
syms theta_ast beta gamma
dct = diff((cos(theta_ast)-beta) / (1-beta*cos(theta_ast)));
dct = simplify(dct);
subs(dct,1-beta^2,1/gamma^2)
```

ans = $-\dfrac{\sin(\theta^*)}{\gamma^2(\beta\cos(\theta^*)-1)^2}$

For relativistic motion there are substantial changes to the angular distribution and the power spectrum. Those occur because acceleration, as defined with local clocks and rulers, is not a quantity which is an inertial frame invariant in contrast to the Lorentz force. Nor does it transform as does the momentum in SR. Expanding to the UR case, the factor $(1 - \beta\cos\theta)$ shows the search light effect that $\theta \sim 1/\gamma$. The exact solution for $\langle U\rangle$ as a function of angle and emitted frequency ω is plotted below. The modified Bessel functions, $K_v(s)$, are available in MATLAB as are almost all the special functions which appear in classical mathematics. The dimensionless distribution shown below is for a circular orbit with circular frequency ω_o. The variables y and s are dimensionless and scaled to be of order 1, so that the photon frequency typically has values which are expected to be greatly increased over the rotational frequency, $\omega \sim \omega_o\gamma^3$. The double differential cross section is dimensionless and the radiative coupling factor α appears as it does for Cerenkov emission, since both involve a single vertex for real photon emission.

The doubly differential cross section is displayed as a surface in (y, s) space where the parameters, ω_o and γ, are set to be those of the LEP collider at CERN with 50 — 50 GeV electron–positron collisions. There are correlations between the angle and the frequency of the radiation. Note that the angular distribution is more peaked at large values of s than at smaller emitted frequencies. At all angles there is a steep falloff of energy with frequencies beyond a cutoff frequency ω_c, defined to be $\omega_o\gamma^3$ which appears in the scaled variable s.

```
% look at full synch dU/domega*frequency
% LEP - 2 parameters energy and rotational freq
fo = 1.1e4; % 11 kHZ
wo = 2 .*pi .*fo;
me = 5.11e-4; % GeV
g = 50 ./me; % LEP at 50 GeV x 50 GeV
        K1 = besselk(2 ./3,s(i,j));
        K2 = besselk(1 ./3,s(i,j));
        dU(i,j) = dU(i,j) .*(K1 .^2 + (y(i) .^2 .*K2 .^2) ./(1.0 +
        y(i) .^2));
```

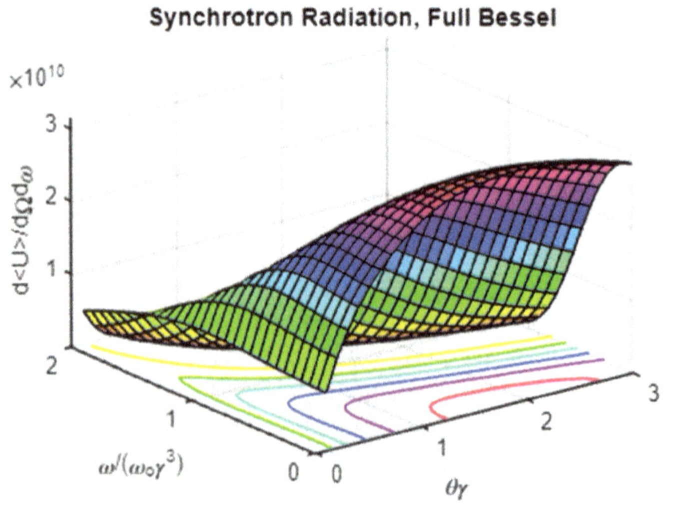

Figure 7.13: Surface of the double differential distribution of radiated energy as a function of the angle and frequency of the emitted photon.

The radiated photon energies extend up to a frequency which is approximately ω_c, much larger than the rotation frequency. The emission angle is $\sim 1/\gamma$ at the highest frequencies, as expected from the search light effect. More generally, $\theta_c = (1/\gamma)(\omega_c/\omega_o)^{1/3}$ and higher energy photons have smaller critical angles as seen above. To convert from power, P to ΔU, the energy loss per circular turn at UR energies, is $\Delta U = P\tau$. The total beam power lost is the energy loss per turn per particle times the beam current in A, approximating UR behavior with $\beta = 1$.

The differential power into unit solid angle is found by integration over s, and can be solved for analytically without the small angle approximation, as shown below. The acceleration here is taken to be $a = (\beta c)^2/\rho$, $m = 1$, using the NR centrifugal force in the case of circular motion. In general, in SR the acceleration is quite a complex quantity, the discussion of which is avoided here. However, the reader should be aware that radiation for linear acceleration differs profoundly from radiation for circular acceleration. A plot is made for the case where $\cos\phi = 0$. The total radiated power is the NR result, Larmor, times $(\beta\gamma)^4$ with $q = 1$. In the UR case the angular distribution of the radiated power is quoted below. The NR Larmor factors appear again, but modified by UR angular factors $1/(\gamma^2(1-\beta\cos\theta)^5)$.

$$
\begin{aligned}
a_1 &= a^2\left(q^2\alpha\hbar/4\pi c^2\right)/(1-\beta\cos\theta)^3 \\
d\langle P\rangle/d\Omega &= a_1(1-\sin^2\theta\cos^2\phi)/[\gamma^2(1-\beta\cos\theta)^2])
\end{aligned}
\tag{7.10}
$$

The radiated photons are very sharply peaked in the direction of the instantaneous electron velocity. Integrating over the solid angle of the emitted photons, the time averaged power is shown below, with $q = -1$ (electrons). The result is the same as the NR Larmor result multiplied by a factor $(\beta\gamma)^4$.

$$
\langle P\rangle = e^2c(\beta\gamma)^4/(6\pi\varepsilon_o\rho^2) = (2/3)\alpha\hbar(c/\rho)^2(\beta\gamma)^4 \tag{7.11}
$$

For UR electrons this factor can be enormous. A hand-waving argument for the factor is that acceleration has dimensions $\{1/t^2\}$, coordinate time in SR scales as $1/\gamma$ and the radiated power scales as a^2, which taken together yields the factor γ^4. This is not a serious argument but it can serve as a mnemonic. Indeed, the large factor does not occur for linear acceleration, along the direction of motion. The expression ΔU is for the radiated energy per turn, $\Delta U = \langle P\rangle\tau$.

$$
\langle P\rangle = (2/3)q^2\alpha\hbar\omega_o^2\gamma^4, \quad \tau = 2\pi/\omega_o, \quad \Delta U = 4\pi\alpha q^2\hbar\omega_o\gamma^4/3 \tag{7.12}
$$

Numerical values for the radiated power, in keV as a function of bend radius for an electron energy of ε are shown below. These quantities are per electron, so that the beam current being accelerated is also a critical, but linear, factor in assessing the total power. This power needs to be supplied to make up for the radiated energy loss so that the electrons remain coasting in the beam pipe

$$\Delta U = (\text{qe})^2\gamma^4/(3\varepsilon_o\rho)$$
$$\Delta U(\text{keV}) = 88.9\varepsilon^4/\rho(\text{GeV}/m) \tag{7.13}$$
$$\hbar\omega_c(\text{keV}) = 2.2\varepsilon^3/\rho(\text{GeV}/m) = 0.665\varepsilon^2 B(\text{GeVT})$$

```
% synchrotron radiation, circular motion only
% angular distribution
```

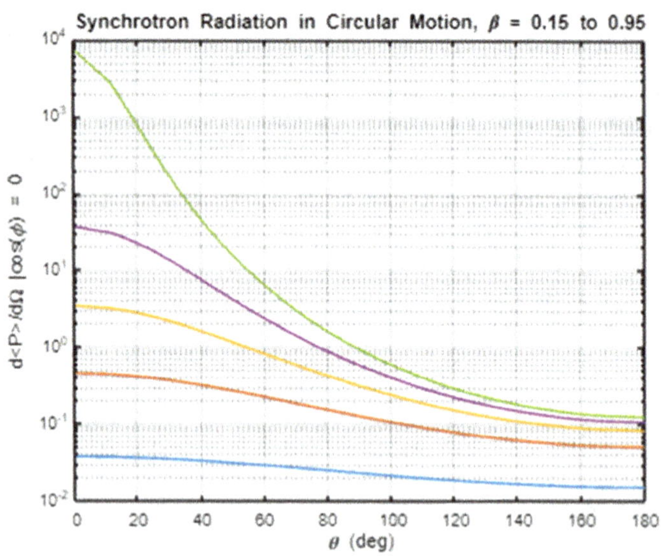

Figure 7.14: Radiated power per unit solid angle as a function of the angle of the emitted photon, for selected values of the electron velocity.

7.9 Steller Fermi Pressure

Stars are composed of mostly hydrogen undergoing fusion reactions which release enormous amounts of energy. The radiation pressure of the photons is balanced by gravitational attraction, and stars can be stable for billions of years. However, ultimately gravity wins. As stars run out of fusion fuel they cool and will begin to contract under their self-gravitational interactions because the radiation pressure can no longer balance the gravitational attraction. For a uniform density star of mass M and radius R the gravitational binding energy U is:

$$U = -(4\pi)^2 G\rho^2 R^5 \sim -\mathrm{GM}^2/R \tag{7.14}$$

This energy is dimensionally $-\mathrm{GM}^2/R$. The pressure, P, is due to compression. There is a change in gravitational self-energy as a function of volume, V, which is the pressure P: It scales as the inverse fourth power of the radius. Expressed as a function of the total number of nucleons N, it scales as N squared.

$$P = \frac{\partial}{\partial V} U \sim \mathrm{GM}^2/R^4 \sim N^2/V^{4/3} \tag{7.15}$$

With all fusion fuel gone, there is nothing to oppose this pressure and avert a collapse of the star except the Fermi pressure of the fermions, the electrons and nucleons, in the star. The gravitational pressure scale is $\sim P_G = \mathrm{GM}^2/R^4$. Previously, looking at the Fermi chemical potential it was found that the Fermi wave vector was $k_F = (3\pi^2 n)^{1/3}$ where n is the number density of fermions. The Fermi energy itself scales as k_F^2 or $n^{2/3}$. The Fermi pressure rises faster than the gravitational pressure, so an equilibrium is possible.

$$U_F \sim \mathrm{NE}_F \sim N^{5/3}/V^{2/3}, \quad P_F = \mathrm{dU}_F/\mathrm{dV} \sim n^{5/3} \tag{7.16}$$

As the density rises with the contraction of the star, the electrons become relativistic and their energy scales only linearly with the momentum rather than quadraticly. At that point the electrons cannot halt the gravitational contracting. The electrons are pushed into the protons and the resulting neutrons continue to resist the contraction while they are non-relativistic. For a total number of nucleons N, a stable radius R_n exists in the balance of gravity with pressure $\sim 1/V^{4/3}$ and the neutron Fermi pressure, $\sim 1/V^{5/3}$.

$$R_n = (81n^2/16)^{1/3}\hbar^2 N^{-1/3}/\mathrm{Gm}_n^3 \tag{7.17}$$

The velocity of particles near the top of the Fermi sea is $\sim \hbar k_F/m$. If the contraction continues the neutrons will become relativistic.

$$\beta \sim (\pi\hbar/\mathrm{mc})(3^{1/6})n^{1/3}/\sqrt{2} \tag{7.18}$$

The gravitational energy scales as N^2 while the fermionic energy scales more weakly with N. When the neutrons become relativistic, $p_F = \hbar \mathrm{c} n^{4/3}$.

A sufficiently massive star, with mass ~ a solar mass, cannot stop the gravitational collapse. The code below graphs the Fermi pressure of electrons and neutrons as a function of R for a star with a user chosen mass. The points are shown only for the regime of NR fermions. The Schwarzschild radius for the Sun is also shown for reference as is the current data point for the Sun. The Schwarzschild radius is the radius where the gravitational force becomes strong and where general relativistic effects are expected to become important, since the potential energy is $\sim Mc^2$ at that radius. The solar radius where collapse cannot be halted is a few km. It seems that for around a solar mass even the neutrons cannot halt a gravitational collapse. At 5 solar masses, the electron and nucleon pressure are always below the gravitational pressure, at all R values. At 1/2 a solar mass the fermions rise above the gravitational pressure and collapse can be averted. The user chooses the mass of the star.

```
% constants, MKS
me = 9.1e-31;
mn = 1.69e-27;
hbar = 1.11e-34;
c = 3.0e8;
G = 6.67e-11;
k = 1.38e-23;
% solar parameters
rsun = 6.96e8 ;    % solar radius , mass and mean density
msun = 2.0e30;
rhosun = (msun .*3.0) ./(4.0 .*pi .*rsun .^3) ; % assumed uniform
Psun = 2.6e16; % solar core pressure in Nt/m^2
%
M = 1.1; %  Mass in solar mass units
rsc = (2.0 .*G .*M) ./(c .^2)     % Schwartzschild radius (m)
```

```
rsc = 3.2609e+03
```

```
N = M ./mn; % number of nucleons
rho = M ./V; %  mass density - uniform
n = rho ./mn; % number density - uniform, all H, Z/A = 1
% evaluate gravitational and e, n  pressure
Pg = (3 .*G .*M ^2)./(8 .*pi .*R .^4); % core pressure - const
density
Rn = (((9 .*N) ./(8 .*pi .*pi)) .^1.66667 .*hbar .*hbar .*pi .*pi
.*pi) ./(15 .*mn);
```

```
Rn = (Rn .*8 .*pi) ./(3 .*G .*M .*M) % radius where n become
relativistic
```

```
Rn = 1.6272e+03
```

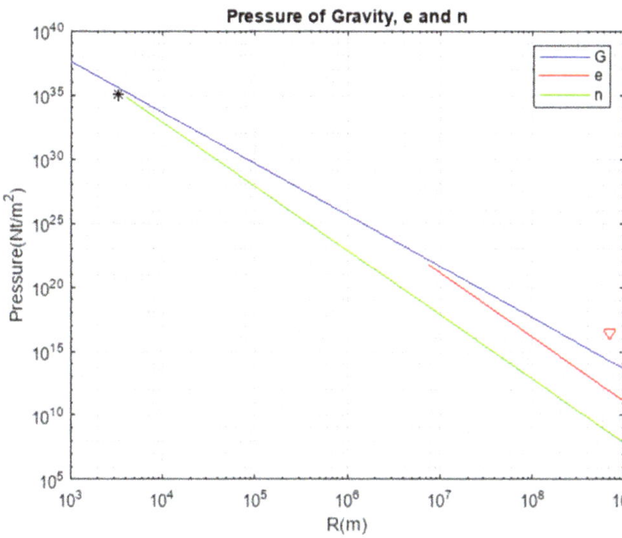

Figure 7.15: Pressure for the Sun as a function of the radius, assuming no fusion is active. The Fermi pressure of the electrons and neutrons are shown until they become relativistic (black * for the n).

7.10 White Dwarf Star

It is possible that the mass of a star is small enough that it can be stabilized against gravitational contraction by the Fermi pressure of the electrons, or as a last resort, the neutrons. As fusion progresses, heavier nuclei are fused. However, beyond the iron nucleus, no further exothermic reactions are possible and the star will cool. All the heavier elements come from supernova explosions, not ordinary fusion reactions. In what follows the assumption is that the star is to be stabilized by Fermi pressure. A simple model was previously made by looking at electron and neutron Fermi pressure and cutting off when the fermions became relativistic. In that simplest model, a uniform density star was assumed, and a mass less than about 2 solar masses could be stabilized. In this script a smooth interpolation for NR to UR is made in an effort to make more realistic model The Fermi momentum is p_F and the interpolation function, y_F goes smoothly from NR Fermi momentum to UR with x^2 to x behavior. The electron to nucleon ratio in the star is Ye $\sim$ 0.5 but <0.5 would obtain assuming iron nuclei as the fusion endpoint.

$$\begin{aligned} n_o &= (m_e c/\hbar)^3/3\pi^2 \\ x &= p_F/(m_e c) = (n/n_o)^{1/3} \\ \gamma_F &\sim x^2/\sqrt{1+x^2} \\ \rho &= (m_p n_o)/Y_e \end{aligned} \tag{7.19}$$

The numerical equations as the electrons become relativistic are solved using "ode45" and are:

$$\begin{aligned} \mathrm{dM}(r)/\mathrm{dr} &\sim r^2\rho(r) \\ d\rho(r)/\mathrm{dr} &\sim [M(r)\rho(r)]/(\gamma_F r^2) \end{aligned} \tag{7.20}$$

The results extend the nucleon pressures a bit more realistically. Nevertheless an iron star, with no radiation pressure cannot be stabilized, in this simple model, if it's mass is greater than ~2.0 solar masses. Such a star must collapse into a singularity, a "black hole". The exact mass depends on the stellar composition. In addition, in GR the pressure is not the classical value assumed so far. This effect will be explored momentarily. More detailed models have been made, beginning with the first mass limit, the Chandrasekhar limit, of 1.4 solar masses. Present models find the limit in the range of 2-3 solar masses. In later discussions, the GR concept of a black hole will be explored in more detail.

```
% Model of White Dwarf - Pressure from e,n Mass from Fe nuclei
% constants, MKS. UR to SR interpolation
c = 3e8; hbar = 1.05e-34; me = 9.1e-31;mp = 1.67e-27; G = 6.67e-11;
Ye = 0.464;    % e/nucleon for Fe star, ~ Z/A in general
% e supply pressure, Fermi gas at T = 0
% nuclei supply mass ~ at rest
% # density in #/m^3 when pf = mec (UR/NR boundary)
no = ((me .*c .*sqrt(2))./(pi .*hbar .*(3 .^0.1666))) .^3
```

```
no = 9.2587e+35
```

```
% star cannot be stable at much larger e densities
% overall charge neutral -> e number density => star mass density
rho = (mp .*no) ./Ye % nucleon mass density at no for e
```

```
rho = 3.3323e+09
```

```
% at this density, e become relativistic and cannot support G
% scale for radius and mass at ne = no in constant denstiy
approximation
Rosq = ((no .^1.666) .*(hbar .^2)) ./(me .*G .*rho .*rho);
Ro = sqrt(Rosq) .*2 ;
Mo = (4 .*pi .*rho .* (Ro .^3)) ./3;
Rsun = 7 .* 10 .^8 ;% sun radius
Msun = 2 .* 10 .^30 ;% sun mass
Rscale = Ro ./Rsun
```

```
Rscale = 0.0105
```

```
Mscale = Mo ./Msun
```

```
Mscale = 2.8046
```

```
%Rs = (2 .*G .*Mo) ./c .^2; % Schwarzschild radius
```

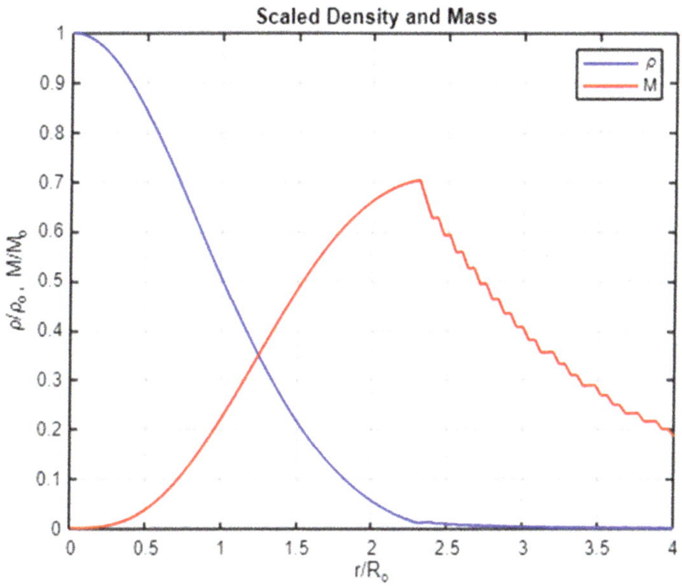

Figure 7.16: Dependence of the scaled density and mass on the scaled radius. The plot is digitally unstable at about 2.4 times the radius of the Sun, 70% of the solar mass.

```
max(MM) % maximum mass for electron stabilized, try neutrons too
```

ans = 1.9707

```
% look at implicit M vs R plot
```

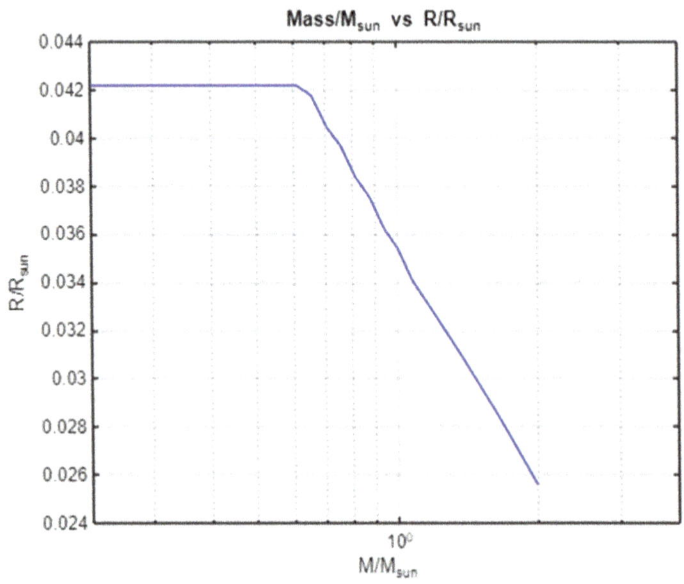

Figure 7.17: Dependence of the radius on the mass of a constant density star. The plot terminates at about 2 times the mass of the Sun, but decreases in radius for masses >0.7 solar masses.

With the smooth momentum interpolation from NR to UR behavior the maximum n stabilized white dwarf mass is still only about 2 solar masses contained in a radius of about 2.4% of the solar radius.

7.11 GR Tests, Red Shift, Light Deflection, Perihelion Advance

There are 3 major advances in physics in the twentieth century; quantum mechanics, SR, GR and cosmology. In GR there are few closed form solutions and superposition is impossible since gravitational energy gravitates.

That means one cannot build up more complex solutions by adding point solutions, as is done in electromagnetism. However, the basic idea is familiar. There is a source term comparable to charge. In this case the source is mass and pressure, ρc^2 and p which are both dimensionally energy densities and elements of the stress-energy tensor, T_{uv}. If a quantum theory of gravity were to emerge in the future, the quanta of the field would have spin $= 2$, similar to the spin 1 quanta of the electromagnetic vector field. The classical field equations are differential equations connecting the source tensor to the space-time metric g_{uv}. Unfortunately, they are non-linear equations and only a few closed form solutions are known. There are 3 well known closed form solutions to the field equations, and the resulting metrics will be examined to see the character of the solutions. In general the Einstein field equations of GR are:

$$R_{\mathrm{uv}}(g_{\mathrm{uv}}) + \Lambda g_{\mathrm{uv}} = (8\pi G/c^4)T_{\mathrm{uv}} \tag{7.21}$$

The R_{uv} is a second rank tensor acting on the metric with differential operators. The term with the parameter Λ, proportional to the metric is an interesting one. It was a term added by Einstein and retracted as what he called his biggest mistake. However, it now has a new life as a "cosmological term" and can describe the "dark energy" which appears to be the majority of the energy in the Universe. The field equations say that the geometry, the left hand side, is caused by the energy content of the space. In what follows only the most simple solutions are explored, and there are only a few such simple solutions. The earliest experimental tests of GR can all be understood by fairly basic arguments. Given a metric, the geodesic equations for the path of a test particle are often fairly easy to solve.

Three early tests of GR had to do with the red shift of photons climbing out of a gravity well (Earth), the deflection of light travelling near the Sun during a solar eclipse, and the perihelion advance of Mercury. In the case of the perihelion advance, the GR geodesic equations added a small force which was not inverse square which therefore made the orbit no longer reentrant. The red shift follows from energy conservation. The photon energy at some height is $\varepsilon = \hbar\omega$. After rising by a height Δy from near the Earth's radius, R_E, the photons lose energy and their frequency is shifted toward the red. The Schwarzschild radius is $R_s = 2\mathrm{GM}/c^2$. The red shift shows that light, having energy, has gravitational mass. A very schematic view of the effect is shown below.

$$\Delta\omega/\omega = -(\Delta y/R_E)(R_s/2R_E) \tag{7.22}$$

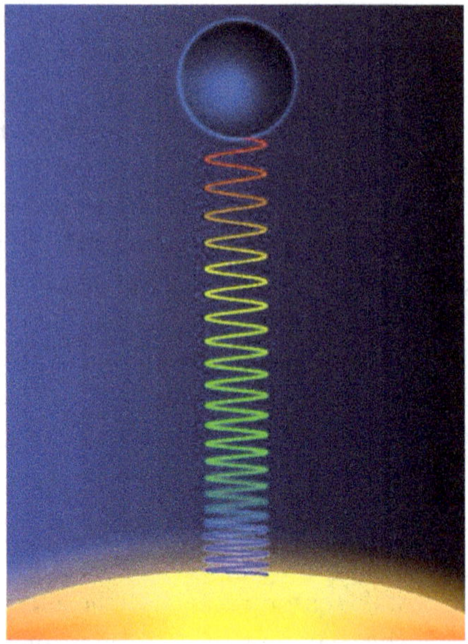

Figure 7.18: Schematic of the gravitational red shift suffered by a photon in climbing out of the gravitational field of the Earth.

Another early test of GR was the observation that light was deflected by passing near the Sun. As with the red shift, light, having mass, was "attracted" to the Sun. The geodesic equations for orbits in a Schwarzschild space will be developed later in a general case as was done for Kepler orbits and then solved numerically. For now, one can simply accept the resulting equation of motion for photons.

$$\frac{\mathrm{d}^2}{\mathrm{d}\theta^2}u + u = 3\mathrm{GMu}^2, \quad u \sim \sin(\theta)/r, \quad 3\mathrm{GMu}^2 \sim 3\mathrm{GM}(\sin\theta/R)^2 \quad (7.23)$$

It has no closed form solution. An approximate solution comes from taking the small term due to GR and using a Newtonian straight line approximation for u. The experimental result for the Sun at eclipse is $1.75''$. In general, $\theta \sim 2R_s/R$, which is small because R_s is a few km, much smaller than the solar radius, R. The user chooses the ratio. The blue line on the plot is an undeflected Newtonian photon. The red line is the approximate GR light deflection for the user chosen value of R/R_s. All 3 early tests of GR were experimentally confirmed. The orbits for light will be explored momentarily.

```
% GR light deflection
% light deflection at min radius R of a body of mass M
MR = 0.02; % Rs/R, point of closest approach
% Newtonian straight line u = 1/r = sin(phi)/R
% GR not in closed form, start with u = 1/r = sin(phi)/R
% Newtonian asymptotes at phi ~ 0, no deflection
% GR asymptotes, small angle approx, at +- Rs/R, total deflect 2Rs/R
R = 1;    % units of solar radius
% Newtonian straight line
% approx GR with curvataure and asymptotes
```

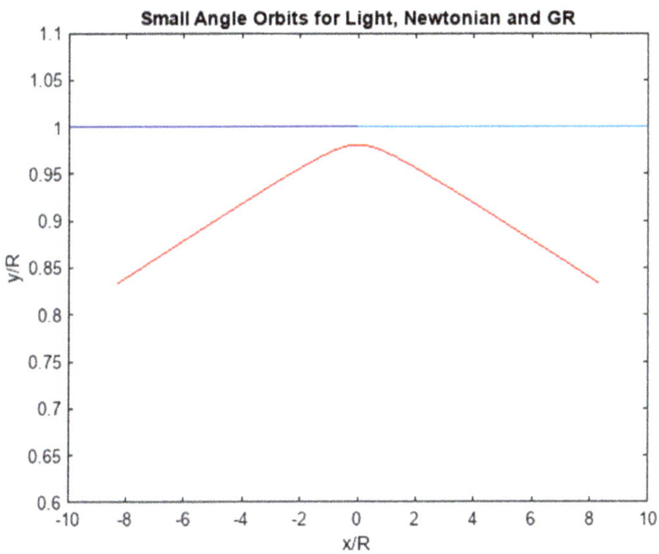

Figure 7.19: Schematic of the gravitational deflection of light when passing near a mass such as the Sun with center of mass at $(0, 0)$.

7.12 Numerical Orbits for Light and Matter

In Newtonian mechanics orbits in time were first explored numerically followed by orbits scaled using energy and angular momentum. The same initial approach is taken here to explore orbits for Newtonian physics, for a material particle in a Schwarzschild metric and for a photon. The parameters that define the geodesic are the initial radius, the initial θ direction of the velocity (set by the initial impact parameter) and the initial velocity, set by the angular momentum parameter which in GR is $h \sim L/c$. The user supplies those parameters as well as choosing Newtonian orbits or GR geodesics for a material particle or light orbits. The material particle GR geodesic equation of motion is:

$$\frac{\mathrm{d}^2}{\mathrm{d}\theta^2}u + u = R_s/2h^2 + 3R_s u^2/2, \quad u = 1/r \tag{7.24}$$

The constant of the motion is now h which is closely related to the angular momentum L; $h = (L/c)/(1 - R_s/r) \sim L/c$, $L = r^2(d\theta/\mathrm{d}t)$. Note the presence of an additional term with respect to the Newtonian case. That means the orbits are no longer defined by an inverse square force law and the orbits will not be re-entrant. This GR term is the origin of the perihelion advance which is the third initial historical test of GR, after the red shift of light and the deflection of light by the Sun. The GR equation for light occurs in the special case where $h \to \infty$ which sets the classical centrifugal force term to 0.

There are a lot of parameters to set and the user is, as always, encouraged to play. However, the classical settings ro = b = L = 5 yield a reentrant ellipse with small eccentricity. On the other hand for GR with ro = b = L = 5 the mass falls into the r = 0 and there is an error message. However, increasing L to 6 shows a nice ellipse with a perihelion advance. For light, reverting to the settings for classical shows a light deflecting, but an unbound, orbit. In the plot the red circle is Rs.

```
% Gen Geodesic - numerical evaluation in a Schwarz and classical
space
rs = (2.0 .*G .*M) ./c .^2;            % Schwarz radius
type = 2; % Classical=1,GR=2,Light=3;
ro = 5; % Enter Initial Distance ro/rs)
roa = ro .*rs ; % in m
vc = sqrt((G .*M) ./roa); % circular velocity
% angular momentum = roa .* vc,   m =1
Lc = roa .*vc;
Lc = Lc ./c; % units of length  since m = 1
% classical eq is d2u/dphi2+u = const = [rs/(L/c)^2]/2
% n.b. L/c is r^2dphi/dt classically, dphi/ds in GR
b = 5 ; % Impact Parameter b in rs units
ba = b .*rs;
sp = ba ./roa; % initial angle
cp = cos(asin(sp));
dudp = cp ./ba;        % initial velocity - assume initial straight
line path
L = 6; % Enter Angular Momentum, L/c (m) = betao*b in km units ');
```

```
L = L * 1000;
cc = (2.0 .*L .*L) ./rs; % classical constant Term
const = 1.0 ./cc;
% GR term
gr = (3.0 .*rs) ./2.0;
% Gr for light has h -> inf
```

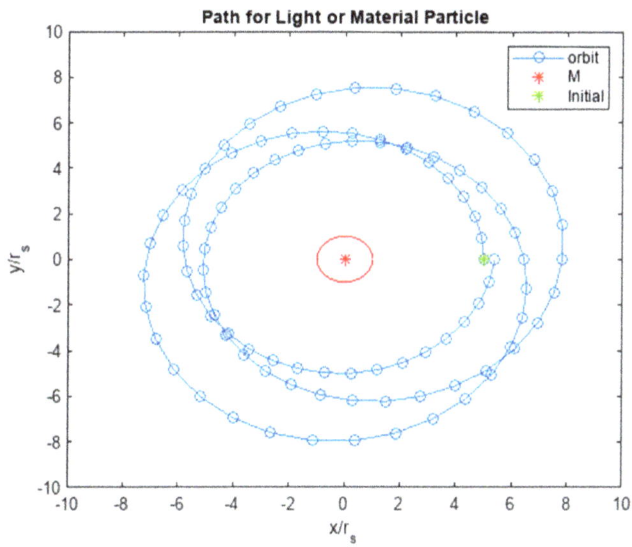

Figure 7.20: Last frame of a movie of a test mass orbiting in a Schwarzschild metric. The Schwarzschild radius is the red circle. The orbit is clearly not re-entrant.

7.13 Radial, Circular Geodesics in a Schwarzschild Space

In Newtonian physics there is a universal time, independent of any relative motion of observers. In SR the constancy of the velocity of light forces the conclusion that clock time is dependent on the relative motion of inertial frames. In SR there is a proper time invariant interval, ds, and a coordinate time dt. The proper time ds is the time for an observer at rest in the inertial frame. The invariance of the interval makes the time dilation effect for dt very visible in SR

$$ds^2 = (cdt)^2 - (\overrightarrow{dx})^2 = (cdt/\gamma)^2 \tag{7.25}$$

Geodesics in a Schwarzschild space are the paths taken by test mass objects in the space which contains large masses which define the metrical shape of the space-time. These paths are the GR analogue of Newtonian orbits classified using energy and angular momentum. Closed form solutions for purely radial and azimuthal (circular) motion are possible. Symbolic integration of the interval is used to solve the geodesic equations in this case. In the GR case, time is also affected by the presence of mass-energy. The simplest case, as in Newtonian physics, is the point mass, or in electromagnetism, the point charge. For a point mass there is a solution of the field equations called the Schwarzschild metric. Indeed, there are very few analytic solutions for the metric in GR. In addition GR is not a linear theory, so that knowing the point solution one cannot superimpose solutions to build up more complex cases. That failure is because the gravitational field has energy and therefore itself gravitates, unlike the electromagnetic field which is not itself charged. The Schwarzschild metric is:

$$\mathrm{ds}^2 = (\mathrm{cdt})^2(1 - R_s/r) - (\mathrm{dr})^2/(1 - R_s/r) - r^2(d\Omega)^2, \quad R_s = (2\mathrm{GM}/c^2) \tag{7.26}$$

The metric is isotropic, but both time and radius are position dependent. The path of a test body, mass, or light energy, in this curved space is called a "geodesic". An example is the great circle route geodesic on the curved surface of the Earth. Something strange happens at the Schwarzschild radius; proper time goes to zero and proper radial distance diverges. One can probe the solution by dropping a small test mass at rest from an initial position. Or one can see if circular orbits exist in this space. In what follows there are 2 definitions of velocity, using either coordinate time or proper time. Using proper time, or time at rest with respect to the traveler, it takes a finite time to fall to the center. For observers using coordinate time, defined as time on clocks far from the mass point, the clocks appear to stop at the Schwarzschild radius.

The test mass is dropped at rest from a radius r_o which value the user can choose. Solutions are found symbolically using the MATLAB utility "int". There are closed form radial geodesics for the radius as a function of proper time or coordinate time. It takes an infinite coordinate time to reach R_s but only a short proper time.

$$(r - R_s)/(r_o - R_s) = e^{-\mathrm{ct}/R_s}, \quad s_o - s = (r^{3/2} - r_o^{3/2})/\sqrt{R_s}. \tag{7.27}$$

```
% radial, circular geodesics in a Schwarzschild space
rss = (2.0 .*G .*M) ./(c .^2);
roo = 3; % ro/rs drop from rest
% Velocities, dr/ds and dr/dct, drds is the classical result
```

```
dsdr = sqrt(r/Rs);
drds = 1/dsdr
```

$$\text{drds} = \frac{1}{\sqrt{\frac{r}{\text{Rs}}}}$$

```
dctdr = sqrt(r/Rs)/(1-Rs/r);
drdct = 1/dctdr
```

$$\text{drdct} = -\frac{\frac{\text{Rs}}{r} - 1}{\sqrt{\frac{r}{\text{Rs}}}}$$

```
% integate over r
```

$$\text{soo} = \frac{2(-\text{ro})^{3/2}\sqrt{-\frac{1}{\text{Rs}}}}{3}$$

```
ct = cto +  int(dctdr,r,ro,r) %  coordinate time, diverges
```

$$\text{ct} = \text{cto} + \int_{\text{ro}}^{r} \left(-\frac{\sqrt{\frac{r}{\text{Rs}}}}{\frac{\text{Rs}}{r} - 1} \right) \mathrm{d}r$$

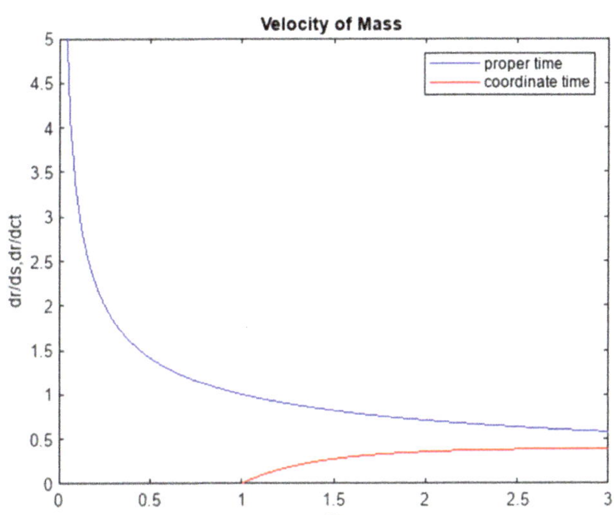

Figure 7.21: Dependence of the velocity of a test mass released at rest and moving radially in a Schwarzschild metric. The velocity in terms of proper time behaves reasonably, but in terms of coordinate time it goes to 0 at the Schwarzschild radius.

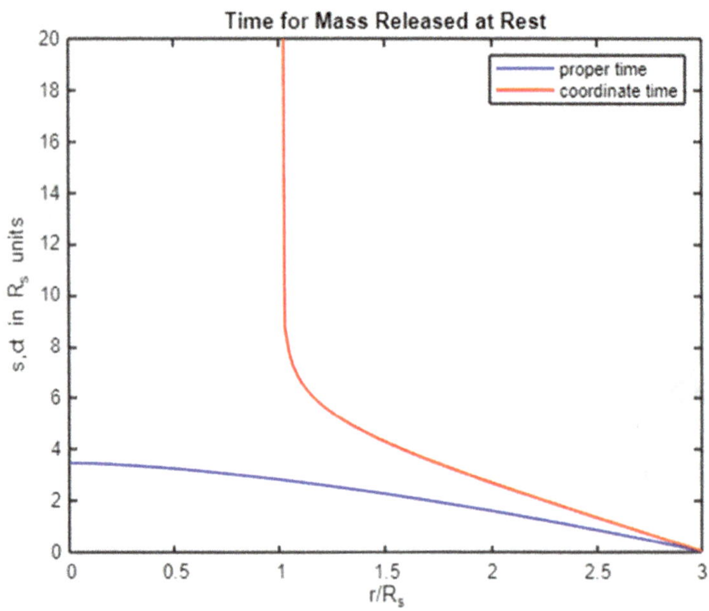

Figure 7.22: Dependence of the coordinate and proper time of a test mass released at rest and moving radially in a Schwarzschild metric as a function of radius. The proper time to reach $r = 0$ is finite, while the coordinate time diverges at the Schwarzschild radius.

One can also solve for circular geodesics. Classically a = the radius = J^2/GM, $J = L/m$, L = mva. For photons the radius is $a_\gamma = 3R_s/2$. for a stable circular orbit. For material particles there is a conserved quantity $h \sim L/c$. The circular radius is $a = h^2/R_s[1 + \sqrt{1 - 3(R_s/h)^2}]$. The constant and radial limits for material particles are $h > \sqrt{3}R_s$, $r > 3R_s$ and no circular orbits are possible at radii less than 3 Schwarzschild radii.

```
% Now the Circular Geodesic
% h is the constant of the motion ~ L/c
```

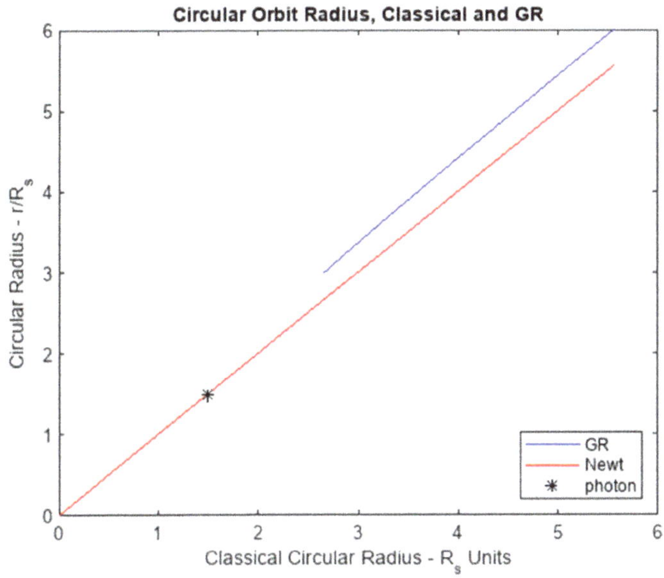

Figure 7.23: Circular orbits in a Schwarzschild space. For Newtonian physics, any radius orbit is possible. For a photon the only possible circular orbit is a 1.5 times the Schwarzschild radius. For a material test particle all the radii >3 Schwarzschild radii are possible circular orbits.

7.14 Orbits in a Schwarzschild or Kerr Metric

There is one other known solution for a point mass, but in this case one with an intrinsic spin, or angular momentum. The Kerr metric has 2 parameters, for mass and spin, M and J. There is as maximum allowed value for J, $J/\mathrm{Mc} = R\beta$ classically, so the constraint $R > J/\mathrm{Mc}$ could be thought of as a velocity of rotation <c. A constraint on the a parameter is that $a < R_s/2$. The Kerr metric is:

$$\begin{aligned} \mathrm{ds}_K^2 &= (\mathrm{cdt})^2(1 - R_s/r) + 2a(R_s/r)(\mathrm{cdt})(d\theta) - (\mathrm{dr}^2)r^2/\Delta \\ &\quad - [r^2 + a^2((1 - R_s/r))]d\theta^2 \\ a &= J/(\mathrm{Mc}), \quad \Delta = r(r - R_s) + a^2 \end{aligned} \tag{7.28}$$

Only equatorial orbits remain planar and only those are explored. There are no purely radial geodesics due to the "frame dragging" of the rotation of the Kerr black hole as seen in the term proportional to $d\theta$(dct). The coupling terms for cdt, $d\theta$ mix the time and angular coordinates which indicates that frame dragging will occur. It can be noted that such cross terms are also seen in Newtonian physics when one transforms to a rotating coordinate system. So such terms are not specific to GR. When $a = 0$, the Kerr metric simplifies and becomes the Schwarzschild metric, as expected. As with the Schwarzschild metric there is a boundary of infinite red shift at $r = R_s$. There is also an equatorial boundary with an event horizon which is also at $r = R_s$, again as for the Schwarzschild case, but now where $\Delta = 0$. There is a region between the 2 boundaries called the ergosphere. The inner boundary is the event horizon while the outer boundary is the is the static limit where coordinate clocks diverge. The locus of the event horizon is:

$$r_{\text{eh}} = R_s[1 + \sqrt{1 - (2a/R_s)^2}]/2, \quad a < R_s/2 \tag{7.29}$$

The geodesic equations are shown explicitly in the code that follows. However they are a bit complex and the user may not wish to delve too deeply. The orbit is plotted and the 'ergosphere boundary is shown in green for the inner limit and black for the outer limit. The user can toggle between Schwarzschild and Kerr solutions and compare them. The geodesic Kerr equations are solved numerically using the utility "ode45".

With the choices E = 1, L = 3, ro = 4, sign = 1 the orbit looks like an unbound Kepler orbit since the location is far enough away from the ergosphere. However, for E = 0.1, L = 1, ro = 4 and sign = −1, the orbit goes into the ergosphere. The inventive user can find much more pathological orbits as desired. The user can easily pick Kerr/Schwarzschild, E, L, ro, and the sense of the rotation.

```
global a E L  ; % constants. spin, mass = 1, energy and angular
momentum
% Kerr metric - geodesics
% specialize to equatorial - remains planer, unique
% input spin - defines event horizons, a is J per unit mass
% G = c = 1 units, fix this parameter, a/M,  M = 1
% event horizon is at - equatorial - M(1+-sqrt(1-(a/M)^2)
% infinite red shift at r=rs. between is ergosphere
% parameters of geodesic eq
so = 0;   % initial proper time
pho = 0;  % initial phi angle
ro = 4; % must be > ergo ~1, Rs units
```

```
% initially incoming or outgoing
sgn = -1;
% Kerr metric has - t, phi independence, -> 2 constants of motion
% analogue of Newtonian orbits
```

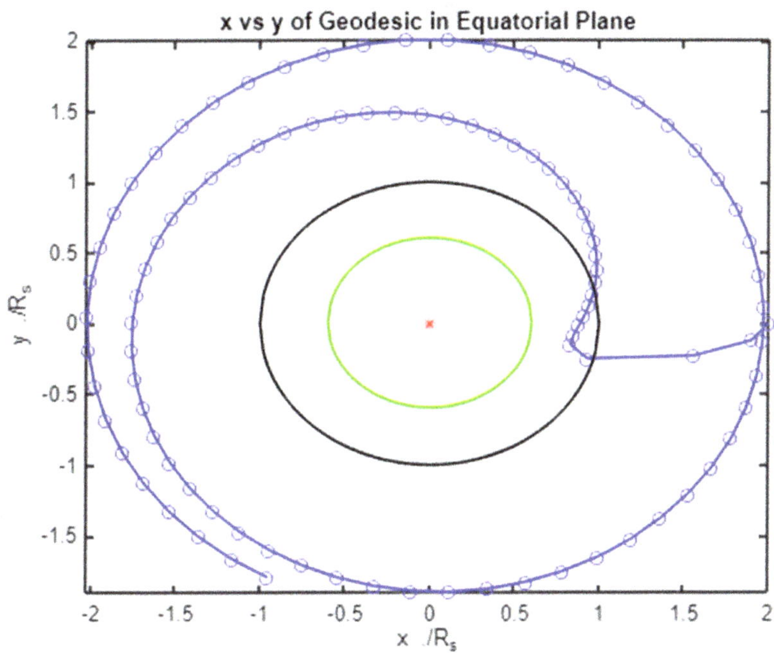

Figure 7.24: A possible trajectory in a Kerr space. The ergosphere, in black, and the Schwarzschild radius, in green, are distinct. In this specific case the frame drag on the test particle is very evident.

7.15 GR and Interior Solar Pressure

The pressure of gravity for the interior of a star is clear. In the uniform density star model of radius R composed entirely of protons, $n = N/V$, $\rho = \mathrm{nm}_p$ the classical Newtonian pressure as a function of r is found by considering the change in pressure with radius r as was done previously in exploring the BVP model for stars.

$$\mathrm{dP}(r)/\mathrm{dr} = -\mathrm{GM}(r)\rho(r)/r^2 \tag{7.30}$$

The interior Newtonian potential is $\Phi = (\mathrm{GM}(r^2 - 3R^2)/2R^3)$. For a constant density star, $M(r) \sim \rho r^3$, $\mathrm{dP}(e)/\mathrm{dr} \sim \rho^2 r$, the solution is simple.

There is a maximum pressure at $r = 0$ and the pressure vanishes at $r = R$, as a boundary condition.

$$P(r) = 2\pi G\rho^2(R^2 - r^2)/3, \quad P(0) = 2\pi G\rho^2 R^2/3, \tag{7.31}$$

The interior Newtonian solution for a star of uniform density scales as $P \sim M^2/R^4$. This result for pressure can be formulated in a way to bring out the connection to GR. First there is a radius where the gravitational energy, GM^2/R is comparable to Mc^2, Define that radius to be $R_s = 2\mathrm{GM}/c^2$, as before. One also expects internal GR effects to be important at radii near R_s. This expression for P has a maximum at $r = 0$, $R_s/4R$. The result for pressure can then be cast into the dimensionless form:

$$P/\rho c^2 = (R_s/R - r^2/R_s^2)/4 \tag{7.32}$$

In GR pressure, with dimension of energy density, appears in the stress energy tensor and is therefore a source term for the metric tensor along with mass. The matter in GR defines the geometry. There is a closed form solution for the metric in the case of a uniform density star called the Oppenheimer-Volkoff solution. It is continuous with the exterior Schwarzschild solution for the metric of a point mass at $r = R$, as it should be. The explicit solution for the interior metric is:

$$\begin{aligned} \mathrm{ds}^2 = &\left[(3/2)\sqrt{1 - R_s/R} - (1/2)\sqrt{1 - (R_s r^2/R^3)}\right]^2 (\mathrm{cdt})^2 \\ &- (\mathrm{dr})^2/(1 - R_s r^2/R^3) + r^2(d\Omega)^2 \end{aligned} \tag{7.33}$$

The expression for the GR dimensionless ratio $P(r)/\rho c^2$ is given in the script below. The script displays the interior pressure for a user defined choice of the R/R_s ratio. Note that for small values of the ratio the effect on pressure is small, while for values near 1 the additional pressure is quite large, many times larger than the Newtonian result.

```
% Newtonian Potential and Pressure
% and GR Pressure for uniform sphere
% x  = P/rhoc^2
x = sqrt(1-(Rs*r^2)/R^3)-sqrt(1-Rs/R);
x = x/(3*sqrt(1-Rs/R)-sqrt(1-(Rs*r^2)/R^3))
```

$$x = -\frac{\sqrt{1-\frac{Rs}{R}}-\sqrt{1-\frac{\mathrm{Rs}\,r^2}{R^3}}}{3\sqrt{1-\frac{\mathrm{Rs}}{R}}-\sqrt{1-\frac{\mathrm{Rs}\,r^2}{R^3}}}$$

```
% Newtonian solutions
phiin = (rs .*(r .*r - 3.0 .* R .*R)) ./(4.0 .*R .^3);
phiout = -rs ./(2.0 .*ro);
```

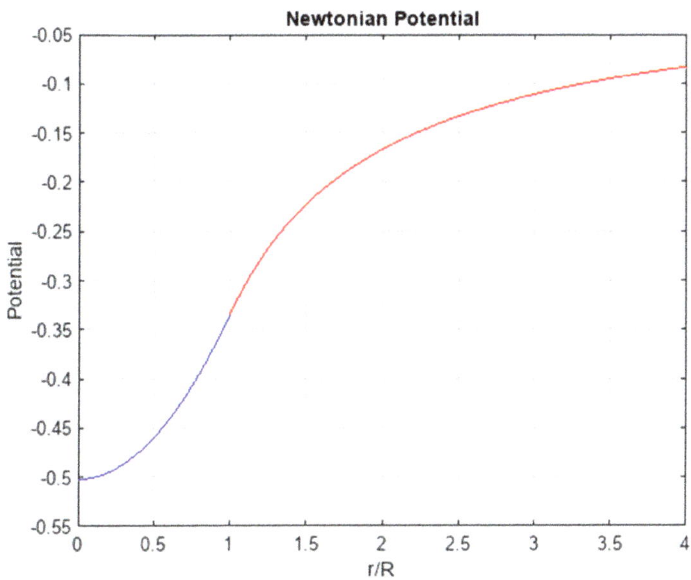

Figure 7.25: Newtonian gravitational potential for a uniform density star.

```
P_rho = (rs - (r .^2).*rs) ./4;
% Max Pressure P(0)/rho*c^2 = rs/4R ;
% now Schwarz GR pressure
% Max GR Pressure P(0)/rho*c^2 = P_rho_S(1));
P_rho_S(1)
```

ans = 0.5883

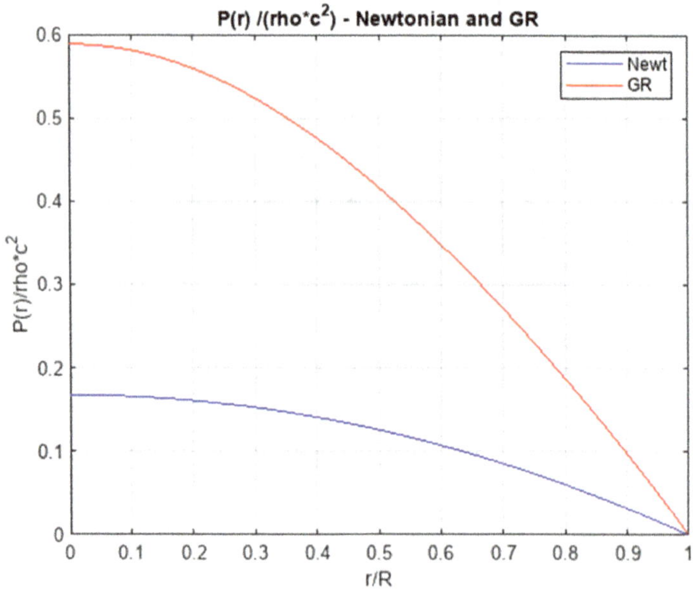

Figure 7.26: Dependence of the scaled pressure interior to a uniform density star for both classical and GR solutions. In extreme cases the GR pressure can be much greater than the classical solution.

7.16 Gravitational Radiation — Binary In-spiral

In electromagnetism an accelerated charge radiates. Mass is the GR "charge". Therefore, an accelerated mass should radiate. The mass defines the metric, so a gravitational wave should be a metric distortion. These effects are not normally obvious to say the least, so how does one establish that gravity waves exist? Since matter (and pressure) are the source terms for the metric in GR, extreme astronomical situations with distance scale near the Schwarzschild radius are places to look for gravitational radiation. A binary star is a candidate source, and will lose energy by radiating gravitational waves and so will in-spiral to near the Schwarzschild radius from an initial orbit. The situation is similar to the problem that atoms are stable, but cannot be classically as they will radiate electromagnetic waves. That problem was solved by quantum mechanics but a quantum theory of gravity is not yet available. In this situation, classical dynamics is applied up to R_s, with an energy loss due to GR radiation. The scale of the maximum wave frequency for solar mass scales is in kHz, hence the name "chirp". In fact an audio file is made by the script using the utility "sound" and executed for the user to hear. A collapse to a black hole is as extreme a situation as can be obtained and it has been shown that massive stars must collapse into such objects. Gravity does win in the end.

The radiation is quadrupole as opposed to the lowest order dipole electromagnetic radiation. There is no negative mass so no dipole whereas charge has both negative and positive constituents. Therefore, the power radiated scales as ω^6 and not the electromagnetic fourth power. Any gravitational system can only radiate up the a maximum power, $P = 2c^5/5G$ which is $1.32 \times 10^{52}\,W$. This is approximately 10^{26} times larger than the solar luminosity, and such a radiation of power should be observable. For the in-spiraling binary, the power is the maximum power times $\sim(R/R_s)^5$. Consider a binary star system rotating about the CM frame at $(0,0)$ with masses $M_1, M_2 = M$ and distances $r_1, r_2 = R$. In the simple case of equal mass stars the rotation frequency ω, power output $\mathrm{dU}_G/\mathrm{dt}$, and the reduction in the period, $d\tau/\mathrm{dt}$ as energy is lost are:

$$\omega^2 = \mathrm{GM}/4R^3$$

$$\mathrm{dU}_G/\mathrm{dt} = (2c^5/5G)(R_s/2R)^5 \tag{7.34}$$

$$d\tau/\mathrm{dt} \sim (R_s/c\tau)^{5/.3} \sim (R_s/R)^{5/2}$$

The following script makes a "movie" of the classical in-spiral and then plots the "chirp" down to a separation of R_s. The dynamics is classical since a GR solution does not exist. The user picks the mass in solar mass units. The user also gets to scale the sound in frequency in order to make it more "chirp" like, since it is only ~kHz. A scale setting of ~2 works fairly well. As a rule of thumb, large objects like whales makes low frequency sounds and vice versa. A MATLAB file of a Handel tune exists and can be scaled up or down for the amusement of the user. A solar scale mass has a Schwarzschild radius ~km which implies radiation with $\lambda \sim R_s$.

```
% Inspiraling Binary due to Grav Radiation, classical binary
dynamics, GR
% energy loss
```

```
mm = 5; % mass in Mo units (1,10)
M = mm .*Mo;
rors = 4.0; % ro in rs units (3,5)
% find the initial orbital frequency and classical time to inspiral
% Initial Orbital Frequency (Hz) = wo
% Classical Inspiral Time to r=0 (sec) = tc
% Schwarzchild Radius (m) = rs
% Max Frequency at rs = ws
% Time to Reach Schwarzchild Radius (sec)= ts
tss = ts .*1000 % inspiral time im msec
```

```
tss = 15.7486
```

```
wss = ws ./1000 % max frequency, at rs, in kHz
```

```
wss = 7.1559
```

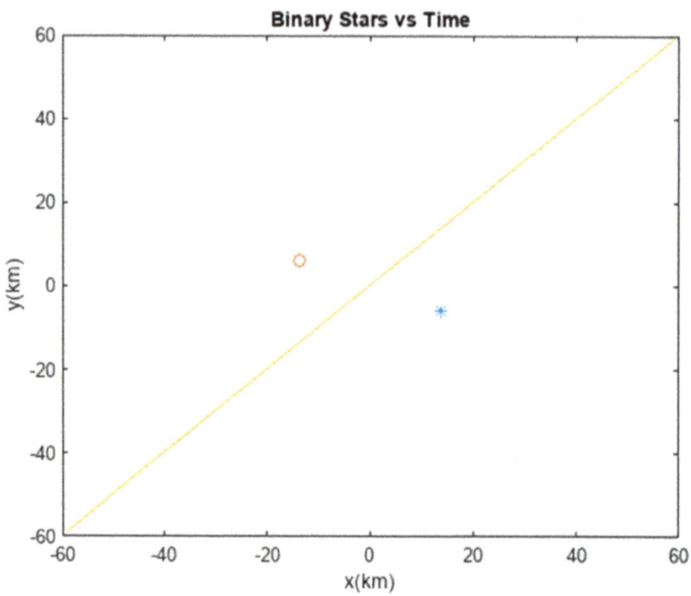

Figure 7.27: Last frame of a movie of the in-spiral of an equal pair mass binary.

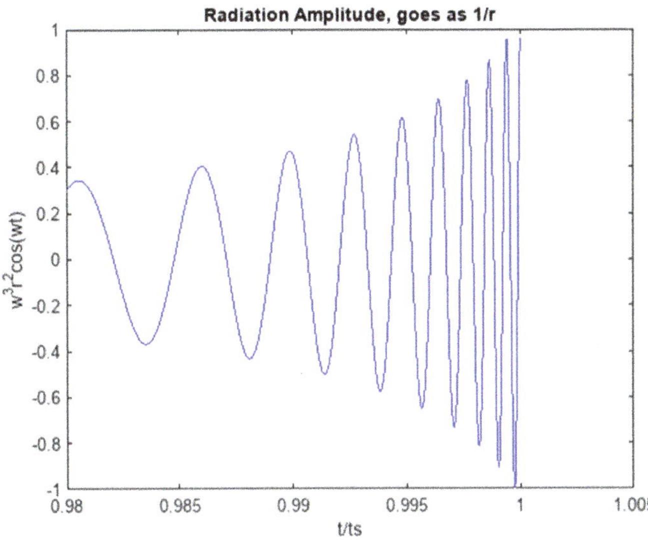

Figure 7.28: Plot of the scaled oscillations of the binary in the last few moments before an equal pair mass binary reaches the Schwarzschild radius.

```
% use MATLAB utilities that cover sound files, a MATLAB Handel file
exists
% choose Handel or chirp
iH = 0
```

```
iH = 0
```

```
load handel.mat
```

7.17 Tidal Forces and Gravity Wave Detection

Gravity can be "removed" by going to a free fall frame — Einstein's Equivalence Principle. You can't tell if there are no forces acting on you or if you are in a free fall frame immersed in a uniform gravity field. However, that is only true at large scale. The tidal components of the force are intrinsic and cannot be transformed away. For an extended object of mass m acted on by a mass M a distance r away the "tidal" forces tend to elongate along the line of centers, z, and compress perpendicular to it, along x. These effects were seen already when the topic of Earth-Moon tides was explored and the tidal deformation of the oceans was displayed.

$$F_z \sim 2z(\mathrm{GMm}/r^3), \quad F_x \sim -x(\mathrm{GMm}/r^3) \tag{7.35}$$

It is these forces of compression and elongation which are driven by gravity waves that are detected in antenna such as the LIGO and VIRGO detectors. A very simplified schematic of the response of these interferometers is produced below. Since the forces are divergence-less a tidal potential can be defined whose gradient is the tidal force, similar to the situation in electromagnetism or fluid flow.

$$\begin{aligned} &V/m = \Phi = -(R_s c^2/2)(z^2 - x^2/2)/(x^2 + z^2)^{3/2} \\ &\vec{F}/m = -\nabla\Phi, \quad F_x/m = -(\mathrm{x}\mathrm{R}_s c^2)/r^3, \quad F_z/m = (\mathrm{z}\mathrm{R}_s c^2)/r^3 \end{aligned} \tag{7.36}$$

The contour of the tidal potential illustrates that the forces are of elongation and compression. The "movie" of the antenna motion as a wave passes by, shows how the deformations of the antenna evolve in time. Note that the movie is very simplified since, as with electromagnetic waves, many states of mixed polarization are possible, It is also notable that a dimensionless metric distortion caused by the wave is, $h \sim R_s^2/(\mathrm{Rr})$ which depends on the source through R_s, R and the observation point, r. Taken together with r an astronomical scale of order >100 ly, the metric distortion is very small and very challenging to detect. The wavelength λ is proportional to R_s and the solar scale for collapse or white dwarf value is a few km. That size defines the size needed for the antenna. Larger objects would require larger antennas, such as the proposed LISA space based array of detectors.

$$h = 8\mathrm{MGR}^2\omega^2/(\mathrm{rc}^4) \sim R_s^2/(\mathrm{Rr}) \tag{7.37}$$

The 2 powers of the Schwarzschild radius mean that these distortions at astronomical distances, r, will be very small and therefore require very sensitive experimental techniques in order to be observable. Nevertheless, such gravitational wave distortions have recently been observed using the km long interferometers LIGO and VIRGO.

```
% Tidal Forces and Gravity, Deformations for a Grav Wave detector
% Tidal Force, Fz = 2*z(GMm)/r^3, Fx = -x(GMm)/r^3, F is
divergenceless, define Tidal Potential
```

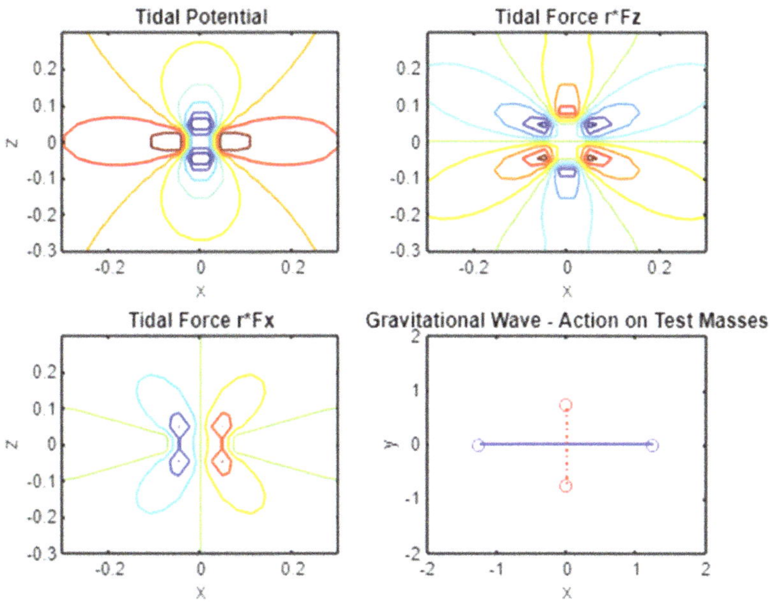

Figure 7.29: Tidal potential (top, left), Tidal force along z (top, right), Tidal force along x (bottom, left) and Gravitational quadrupole antenna (bottom, right).

```
% now look at test mass acceleration
% Gravity Wave - Response of 4 Test Masses
```

Figure 7.30: Photo of the LIGO gravitational wave antenna, a quadrupole interferometer.

Chapter 8

Astrophysics and Cosmology

"And yet it moves" Galileo Galilei

"I ask you to look both ways. For the road to a knowledge of the stars leads through the atom; and important knowledge of the atom has been reached through the stars." Sir Arthur Eddington

"In the beginning there was nothing, which exploded." Terry Pratchett

The last Section of the text is quite short, but this does in no way mean the topics are not relevant and of great interest. Rather, cosmology, in particular, is such a rapidly expanding field that it can only be sampled. Specifically, the present "Standard Model" of cosmology has a situation where $\sim$95% of all the mass in the Universe is taken by "dark energy" and "dark mass". These labels have no known candidates within the scope of the "Standard Model" of fundamental particle physics, with fermions (quarks and leptons) and bosons (force carriers — photons, W, Z bosons, and gluons). The dark objects are "dark" and do not interact via the strong or electromagnetic interactions, and perhaps not with the weak interactions either. Very active searches for dark matter are ongoing, but to date without success in observing these dark objects except by their gravitational effects. The candidates for dark energy are even less clear, only the knowledge that it acts like the vacuum itself within experimental limits.

That being the case, this Section begins with solar system topics and models of the suns. These topic are well understood and indeed, the light element abundance of the Sun was an important input to the emerging cosmological model.

8.1 Solar Exploration — Transfer Orbits

Consider a trip to a nearby planet, such as Mars as a first step in planetary exploration. What rocketry does that entail? The basic idea is to exit the Earth orbit, assumed to be circular, enter an eccentric, elliptical orbit with a velocity change Δv_1, and then enter the orbit of Mars with a second burn of Δv_2 to go from an elliptical Earth orbit to a circular Mars orbit, where the small eccentricity of the orbits is ignored in the interest of simplicity. This "Hohmann" orbit is an elliptical orbit to transfer between two circular orbits. This plan is popular because it uses the minimum amount of energy. The launch is at the Earth orbit and the arrival is at the location of Mars 180 degrees away from Earth at launch. This launch window is not always available and the time must be chosen. The trip takes about 8.5 months.

The analysis of the transfer orbits follows from the general Kepler analysis seen previously in the Mechanics Section. The period of an orbit depends only on the 3/2 power of the semi-major axis, a. The energy, E, of the orbit is a constant of the motion and depends only on the inverse of the elliptical focus a.

$$
\begin{aligned}
E/m &= v^2/2 - \mathrm{GM}/r = -\mathrm{GM}/(2a) \\
v^2 &= \mathrm{GM}(2/r - 1/a) \\
(v/c)^2 &= R_s(1/r - 1/2a)
\end{aligned}
\tag{8.1}
$$

From these results one can find the velocity difference needed to go from a circular orbit with radius R_1 to an elliptical orbit for semi-major axis $R_1 + R_2$. The velocity difference to then inject into a circular orbit at R_2 and the time it takes, the half period of the transfer ellipse, τ are:

$$
\begin{aligned}
&R_1 \rightarrow (R_1 + R_2) \\
&\Delta v_1 = \sqrt{\mathrm{GM}/R_1}(-1 + \sqrt{[2R_2/(R_1 + R_2)]}) \\
&\Delta v_2 = \sqrt{\mathrm{GM}/R_2}(+1 - \sqrt{[2R_1/(R_1 + R_2)]}) \\
&\tau = \pi\sqrt{(R_1 + R_2)^3/\mathrm{GM}}
\end{aligned}
\tag{8.2}
$$

The trip requires two "burns", Δv_1 to go from the velocity at the Earth's radius, R_1, to the Hohmann ellipse and one, Δv_2 to go from the ellipse a half period later into the target orbit, R_2. A one way trip varies from 0.71 years to Mars, 16 years for Uranus, and 31 years to Neptune. Given the radiation issues and the sheer time scale, these trips appear to be limited, at least for human passengers, to Mars at most. The destination planet in the script is chosen by the user. The period, orbital velocity and radius for the chosen planet are displayed. The transfer orbit semi-major axis, the trip time and the two velocity changes are also shown. For a Mars trip the change in velocities at the beginning and end of the trip are only about 10% of the basic orbital velocity. Other, faster, transfer orbits are possible. They would minimize the radiation exposure to cosmic rays. The tradeoff is in the reduction of the mission payload fraction.

```
% Transfer Orbits from Earth to Outer Planets
% Minimum energy Hohmann orbits
% plot for Earth and Mars (first step)
% Me V E Ma J S U N - 8 planets
% circular velocity in km/sec
ass = [0.39 0.73 1.0 1.52 5.2 9.5 19.2 30.1] ;   % circular orbit
radius in Au
tauss = (ass ./ass(3)) .^1.5 ;   % period scale as radius^3/2 in
years
ipl = 2; % 2,3,4,5,6 = Mars,Jupiter, Saturn, Uranus, Neptune
% period (yr)
```

```
ans = 1.8740
```

```
% velocity km/sec
```

```
ans = 24.1000
```

```
% a in AU
```

```
ans = 1.5200
```

```
% trip ion transfer orbit (yr) % half elliptical period
```

```
Trip = 0.7072
```

```
v1 = Vorb(1)   % Earth
```

```
v1 = 29.8000
```

```
v2 = Vorb(ipl) % planet
```

```
v2 = 24.1000
```

```
% find burn velocities for insertion and exit (km/sec)
```

```
dv1 = 2.9305
dv2 = 2.6300
```

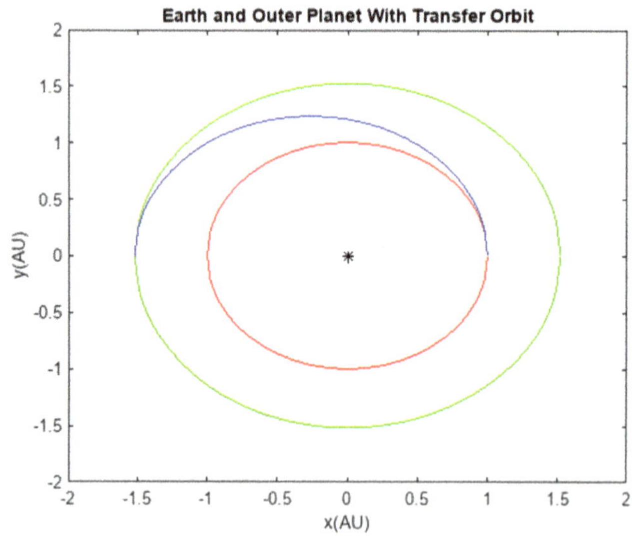

Figure 8.1: Orbits of the Earth and Mars and the transfer orbit between them.

8.2 Solar Sailing

A staple of science fiction is the use of solar sails to cruise around the solar system or even to travel to nearby stars. Light has energy, E, and thus momentum, E/c, and imparts twice that momentum when reflected from a surface. A simple layout is to release such a sail from an initial distance from the Sun and see how much velocity the sail ends up with. The assumption is that the sail is unfurled at a radius r_o very rapidly and then the Sun is used as a launching light source. No laser launchers are used.

The acceleration toward the Sun is just $a_s = \mathrm{GM}(\rho\pi r_s^2 t)/r_o^2$ where the sail (no payload) has density, radius and thickness of ρ, r_s and t. The acceleration away from the sun is due to the reflection of the solar light by the sail, $a_{\text{light}} = 2L_o/[4\pi c\rho t r_o^2]$. Depending on the parameter values, independent of the radius, the sail will either fall into the Sun or be accelerated outward with an ever decreasing acceleration as the attraction of the Sun and the "repulsion" due to light reflection both fall off with distance from the Sun. The results are independent of the radius of the sail since the weight goes as the square of the radius but the number of photons reflected also goes as the square.

By varying the parameters one can see that starting as close to the Sun as possible gives the maximum final velocity. It is also clear that a trip to even the nearest stars is still a lengthy proposition. A final constant terminal velocity is reached fairly quickly as is seen in the movie of the trajectory of the sail. The user picks the initial radius from the Sun in AU and the thickness of the sail, in μm. The solution is found numerically using "ods45". One could also appeal to energy conservation in order to find the terminal velocity.

```
% solar sailing using the momentum of light, U = pc
Lo = 3.9e26  ;    % solar luminosity in Watts
au = 1.5e11 ;  % 1 au in m
Mo = 1.9e30 ;  % sun's mass in kg
ro = 0.45; % drop from initial radius (AU)
st = 0.5; % sail thickness(Mylar + reflector ~ all mylar) in um
% sail has density of mylar (1.39)+ thin Al(2.7)
asol = (G .*Mo) ./(ro .^2);
% asun - asol; % net acceleration
del = (asun - asol) .*ro .*ro;    % constant for accel which goes as
1/r^2
```

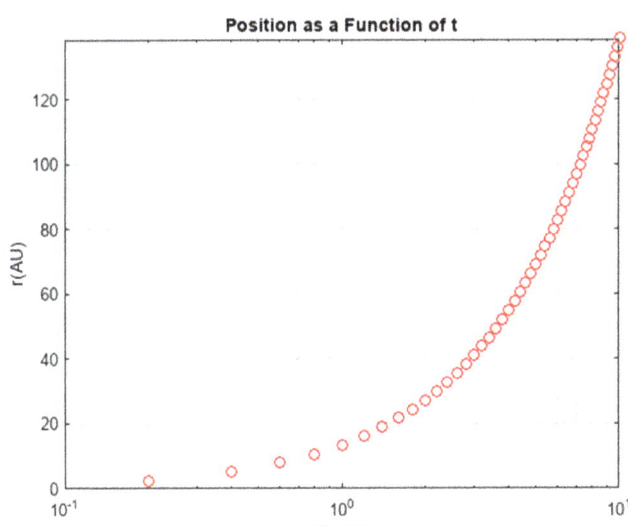

Figure 8.2: Distance covered with a solar sail as a function of travel time.

8.3 Lagrange Points

There has been so far only a look at the gravitational response of 2 bodies to one another. There is clearly a 3 body problem to be explored, but that is intractable and requires numerical solutions. One special case is that of 2 heavy bodies and one very light test body that can be more easily explored. Lagrange points are the stable locations of the light body with respect to the less massive of the 2 bodies. There are 5 stable points, and 3 are collinear with the line between the 2 heavy bodies. Satellites and space based telescopes are placed at these points in order to have a stable observation location. The CM is at $(0, 0)$ which is near the more massive body. The large mass is at $(-x_1, 0)$ while the small mass is at $(x_2, 0)$. Systems presently in use are for the Sun and Earth or the Earth and the Moon.

The mass M_2 is at a distance R from the lighter mass M_1. First, for a body between the two and at a distance r from M_2, which partially cancels the attraction of the M_1 and therefore the period of the test body increases and matches that of the M_2 so that the body is fixed in space with respect to the M_2. Conversely, if the body is further out in radius than M_2 the lighter mass adds attractive force to the test body, which speeds up and stays in synch with the orbit of the M_2. That explains the operation of 2 of the collinear Lagrange points. The third is away from the large mass, opposite to the line of center of the 2 heavy bodies.

In general the velocity goes as the inverse square root of the distance from the mass while the period goes as the 3/2 power, as already seen. The stable points can be evaluated in a co-rotating system where the two heavy masses are at fixed points on the x axis. $x = x_1 + x_2$, $M = M_1 + M_2$, $\omega^2 = \mathrm{GM}/a^3$. A more visual way to see the Lagrange points is to use a potential and look for the potential minima. The potential due to M_1, at $(-x_1, 0)$, and M_2 at $(x_2, 0)$, the center of momentum, CM, at $(0, 0)$ and the centrifugal potential, $\Phi = V/m$ of the orbit about the CM are:

$$\begin{aligned}
&x = x_1 + x_2, \quad M = M_1 + M_2, \quad \omega^2 = \mathrm{GM}/a^3 \\
&x_1 = M_2/M, \quad x_2 = M_1/M, \quad a = x_1 + x_2 \\
&\Phi(x, y) = -G(M_1/R_1 + M_2/R_2) - (\omega R)^2/2 \\
&R_1^2 = (x + x_1)^2 + y^2, \quad R_2^2 = (x - x_2)^2 + y^2
\end{aligned} \tag{8.3}$$

The centrifugal potential scales as radius squared, as seen previously. The masses M_1, M_2 are at distances R_1, R_2 from the observation point, (x, y), located at the test mass.

In the code, the user can choose the mass ratio of the 2 heavy bodies. Small values make the equipotentials more visual. Only the 3 collinear points are explicitly determined. A plot of the potential and the force along the x axis which is the line of centers of the 2 masses is made. The locations of 3 points are visible as the red stars where the force along x vanishes. The locations of L4 and L5 which are at non-zero values of y are not as obvious and have not been well established here. However, for the non-collinear Lagrange points a minimum does exist off the x axis which can be seen for low values of the mass ratio. The MATLAB utility "gradient" applied to the potential is used to find the forces in x and y. The 3 points are found using the utility "sort' on |Fx|. Multiple points are plotted since the "sort" is not very accurate on the Fx array when it is near the zero value at the L2 and L3 locations.

For the real Earth-Moon system, with $M_E/M_M = 82.2$, the L1 and L2 points are about 15% more or less distant than the Earth-Moon distance. For the Sun-Earth system the distances are only about 1% different owing to the larger mass ratio. Many observatories such as Planck and WMAP have been located at L2 for the Sun-Earth system. The Lagrange points are also natural places to find asteroids, for example the "Trojan" L4, L5 points of the Sun-Jupiter system.

```
% 3 body problem, Earth-Moon and Lagrange Points
% Earth-Moon System, Stable Points for a Satellite
% Examples from NASA and ESA are WMAP and PLANCK
% Balance of Gravity and Centrifugal Force
% reference frame rotates with the ~ circular orbital period
% masses are fixed in this frame
   psi(i,j)  = -qq ./((1+qq) .*ree) - 1 ./((1+qq) .*rmm) - rrs ./2;
```

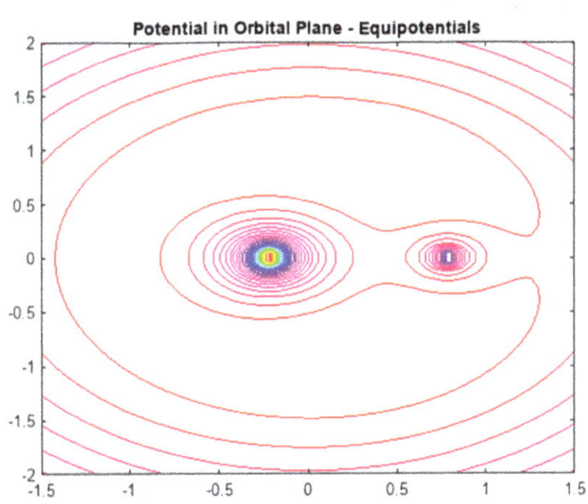

Figure 8.3: Equipotential contours for the Lagrange points.

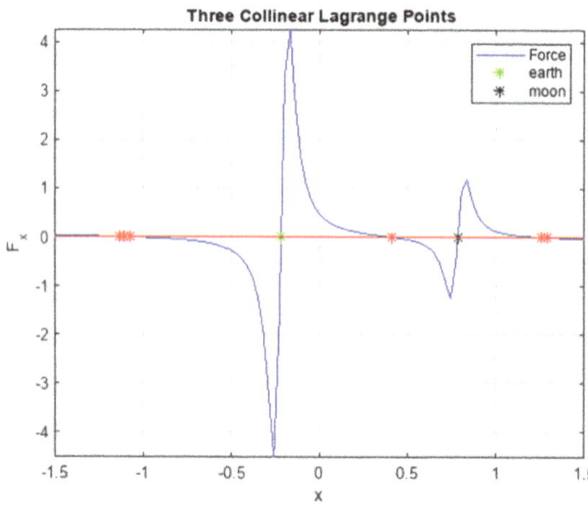

Figure 8.4: Radial force for the Lagrange Points. The red stars are minima found by MATLAB.

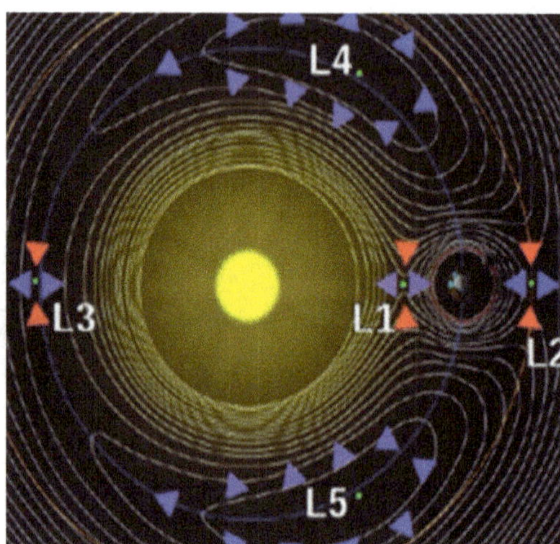

Figure 8.5: Schematic picture of the potential contours and the 5 Lagrange points, where the arrows indicates the balance of forces. The L4 and L5 non-collinear points are clearly seen.

8.4 Solar Model as a Boundary Value Problem

Until the twentieth century the Sun was a problem because the solar energy released could not be explained. The geologists and evolutionary biologists felt that the Earth must have existed for very long times. The physicists did not see how that was possible. It was only with the advent of nuclear physics that the fusion processes which drive the Sun were understood and a lifetime of the sun of billions of years became understandable. There are many models of the sun. A simple approach is adopted here and the results compared to a reasonably successful model which fits solar data. The aim here is to go beyond the uniform density model which has been used so far. To begin, the Sun is treated as an ideal gas with the majority constituents of its mass being protons and Helium ions. The Hydrogen and Helium are ionized at core temperatures corresponding to MeV energies compared to the eV ionization energies. The core is then a neutral plasma of NR protons, electrons and Helium nuclei. The energy density is the product of the number density times the mean thermal energy, (3kT)/2 — the Boltzmann result for the mean.

$$
\begin{aligned}
&\mathrm{PV} = \mathrm{NkT}, \quad n = N/V \\
&m\langle v^2\rangle/2 = 3\mathrm{kT}/2 \\
&P = \rho\mathrm{kT}/(\mu m_p) + 4\sigma T^4/3c
\end{aligned}
\tag{8.4}
$$

In general, there are contributions to the pressure from the gas and from the radiation. The radiation pressure follows from the Planck distribution for energy. The Stefan-Boltzmann constant is $\sigma = 5.67 \times 10^{-8} W/m^2K^o$. The phase space for photons yields an energy density factor x^3 times the Bose-Einstein weight factor. The integrated radiation energy density scales as T^4.

$$
\begin{aligned}
&\mathrm{du}(T)/\mathrm{dE} = (\mathrm{kT})^3/[\pi^2(\hbar c)^3]\{x^3/(e^x - 1)\}, \quad x = E/\mathrm{kT} \\
&u = 4(\sigma T^4)/c = (\mathrm{kT})^4\pi^2/[15(\hbar c)^3]
\end{aligned}
\tag{8.5}
$$

The mass density is related to the number density by μm_p where μ is the mean molecular weight. The mean molecular weight is the average mass of a particle in the gas, so that $\mu m_p \sim 0.62$ for the mix of protons and helium nuclei of a young star when fully ionized, $1/\mu \sim 2\mathrm{X} + 3\mathrm{Y}/4$ where X is the proton mass fraction and Y is the Helium mass fraction. The mean mass is less than a proton mass because of the existence of three electrons in the ionized mixture. The neutral fraction when the gas is not ionized is $1/\mu \sim \mathrm{X} + \mathrm{Y}/4 = 1/1.3$ larger than the proton mass due to the heavier Helium with A of four.

The simplest solar model assumes a steady state stable solution for the star which is characterized by a density ρ, a pressure, P, and a temperature, T, all functions of the radius r from the solar center. The star is in equilibrium between the gravitational pressure and the outward pressure due to the gas and the radiation. In the conventional notation used here $M(r)$, $\rho(r)$, $P(r)$, $T(r)$, denotes the mass within r and the density, pressure and temperature at r. A simplification is to assume that the pressure can be dominated by the contribution due to the gas or to the radiation.

The core where fusion occurs is only about 10% of the mass of the Sun. In fusing protons into helium only about 0.007 of the rest mass is converted to energy. Typical binding energies are 7 MeV while typical light nuclear masses are 1000 MeV, ratio = 0.007. An estimate of the total possible energy release is then $0.1 \times 0.007 \times M_o c^2$ or about $1.3 \times 10^{44}\,J$. Comparing to the solar luminosity, $3.8 \times 10^{26}\,W$. A crude estimate of the solar lifetime is then about 10 billion years. The Sun is expected to burn stably for a good fraction of the time the Universe has been in existence, currently estimated to be about 13 billion years. This simple order of magnitude estimate allows for a long lifetime for the Sun once the fusion process is understood.

The basic equations for the simple solar model, ignoring the radiation pressure in the interest of simplicity, and the boundary conditions, are:

$$\begin{aligned}
&\mathrm{dM}(r)/\mathrm{dr} = 4\pi r^2 \rho(r) && M(0) = 0, \quad M(R) = M_T \\
&\mathrm{dP}(r)/\mathrm{dr} = -\mathrm{GM}(r)\rho(r)/r^2 && P(R) = T(R) = \rho(R) = 0, \\
&P(r) = \rho(r)\mathrm{kT}(r)/\mu m_p &&
\end{aligned} \tag{8.6}$$

The radiation pressure term could be substituted for the temperature term for the pressure, to get a feeling for the relative contribution. The equation for the mass r dependence on density is clear, while the dependence of pressure on G, M and density was mentioned previously, $\mathrm{dP}/\mathrm{dr} \sim G\rho^2 r$, $\rho \sim M/r^3$. The temperature dependence follows from the ideal gas law for pure protons and is altered for a mixture of ions. The boundary conditions at the core and the surface define the meaning here of solar radius, R. The quantity μ is the mean molecular weight, $\mu = \langle m \rangle / m_p$.

The specific heat is defined to be the amount of heat needed to raise the temperature of an object one degree. If the processes are adiabatic, then there is no net heat flow. Rising gas expands and cools. Falling gas contracts and heats. An adiabatic process has PV^γ constant. The factor γ is the ratio of dP/P divided by dV/V which is the ratio of the specific heats and has a value 5/3 for an ideal gas. This is the case for simple convective dominance.

In the case of radiative dominance the photons scatter off the charged protons, electrons and H ions. The scattering is characterized by an opacity, κ, and a luminosity, $L(r)$. For a density ρ and path length z, the photon intensity falls exponentially with z with an exponent of $-\kappa\rho z$. The mean free path is $n\sigma = \kappa\rho$ where σ is the cross section for light scattering off the solar medium. The convective and radiative temperature contributions are:

$$\begin{aligned} \mathrm{dT}(r)/\mathrm{dr} &= \{(\gamma - 1)/\gamma\}\mu m_p \mathrm{GM}(r)/r^2 \\ \mathrm{dT}(r)/\mathrm{dr} &= -3\langle\kappa\rangle\rho(L(r)/4\pi r^2)/16\sigma T^3 \end{aligned} \tag{8.7}$$

The luminosity behavior, $\mathrm{dL}(r)/\mathrm{dr}$ follows that of the mass, but modified by an energy production factor, ε, which, since it represents fusion, is highly temperature dependent and thus biased toward the inner, high temperature core. The opacity, with dimension m^2/kg, is set to a constant times ρ divided by $T^{-3.5}$. The fusion energy production factor is taken to be that for p-p reactions. For the Sun the luminosity to mass ratio implies a mean power production of about 0.2 mW/kg. At the core the power production is the highest, about 4.3 MW/kg due to a core number density of protons of about $9 \times 10^{31}/m^3$.

$$\mathrm{dL}(r)/\mathrm{dr} = 4\pi r^2 \rho(r)\varepsilon = [\mathrm{dM}(r)/\mathrm{dr}]\varepsilon \tag{8.8}$$

Global solar properties are shown in Appendix 9.5. The differential equations are solved as a BVP using the MATLAB utility "bvp4c" which uses the boundary values to solve for the interior parameters in the purely radiative case. The purely convective case is a poorer fit to the "data". Applying the boundary values defines the core values of pressure and temperature. For the Sun the fusion occurs in about the first inner ~30% of the radius, where radiation pressure dominates. Indeed, the convective region for the Sun is only the outermost radii, near the solar surface. The other method of assuming a core pressure and temperature and integrating to the solar surface is not very robust, since the problem is not really an initial value problem, but a BVP. The BVP is not fully stable and the calculation is slow, so be patient. The result is a much improved representation of the Sun compared to trying initial value solutions, although this particular formulation, about as simple as can be and still be able to be a reasonable match to the "data", is slightly unstable.

Scaled variables are used to reduce the variations. They are m, x, p and t defined as: $M = \mathrm{mM}_T$, $r = \mathrm{xR}$, $P = p(GM_T^2/4\pi R^4)$, $T = t(\mu m_p/k)(\mathrm{GM}_T/R)$. The equations are solved for m, p, t, l while ρ is derived using the ideal gas law. Reasonable agreement of the "data" with the model, is found with the BVP approach. This BVP approach is much slower and less robust as the more typical initial value problem solver, "ode45". However, it arrives at a better representation of the solar "data".

```
% treat sun as a BVP -  opacity included - sun
% the set of 4 scaled eqs for m,p,l and t are used and rho derived
from ideal gas
% constants
c = 3e8; % light veocity
G = 6.67e-11 ; % newton constant
mp = 1.67e-27 ; % proton mass kg
sig = 5.67e-8 ; % S-B in W/(m^2deg^4), sig/c is an energy
density/T^4
kb = 1.38e-23 ; % Boltz const J/deg
Ro = 6.96e8 ; % sun radius m
Mo = 2e30 ; % sun mass kgm
Po = 1.29e16 ; %sun central pressure nt/m^2
To = 1.57e7 ; % sun central temp deg K
Lo = 3.8e26 ; % luminosity in W
Teff = 5770 ;  % surface temp , Lo =4*pi*Ro^2*sigma*Teff^4
rhoo = 1.58e5  ; % central sun density kg/m^3
% Solar model from Schwarzchild - assumed to be a good model
% star defined by total mass and radius and bc - need for lumi
R = Ro ; % specialize to the sun
M = Mo;
% dimensionfull scales for dimensionless param m, p, t, l
Pscale = (G .*M .^2) ./(R .^4 .*4 .*pi);
mu = 0.62; % hydrogen by weight, H + He - ionized
Tscale = (mu .*mp .*G .*M) ./(kb .*R);
rhoscale = (mu .*Pscale .*mp) ./(kb .*Tscale); % derived from
ideal gas
% rho = rhoscale(p/t)
Lscale = Lo ; %(G .*M .^2 .*c) ./R .^2;
beta = -(3.0 .*Lscale .*Pscale .*mu .*mp) ./(64 .*pi .*Tscale .^5
.*R .*kb .*sig);
% coefficient in dt/dx = beta*(p*l)/(x^2t^4)*beta in radiative
needs kappa
% evaluate the scaled solution
mu = 0.62; % hydrogen by weight, H + He
% for dt/dx need opacity, density, L and T
% for dl/dx need eps
% dL/dr = eps*rho*4*pi*r^2
```

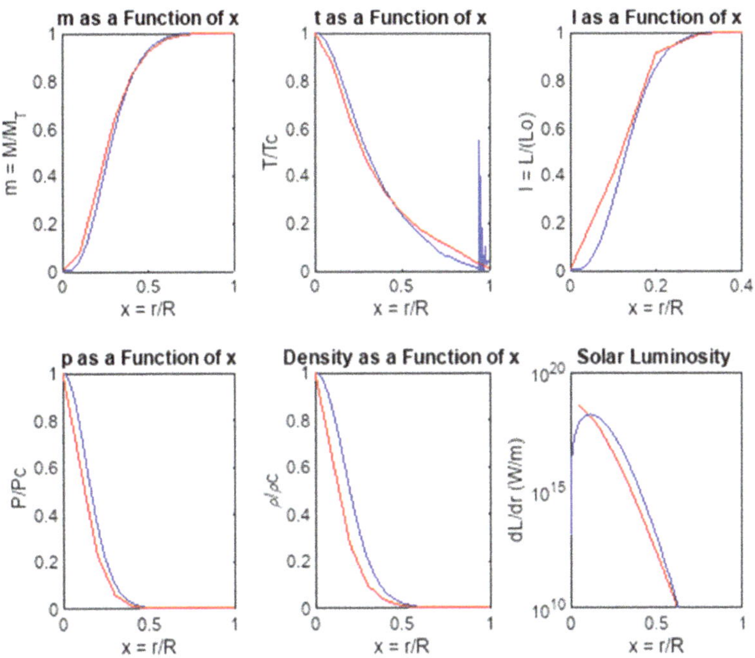

Figure 8.6: Comparison of the results of a successful solar model (red) and the MATLAB BVP solution for mass, temperature, scaled luminosity, pressure, density and solar luminosity as a function of radius (blue).

8.5 Robertson-Walker Metric, Expansion, Power Laws

As we will see, something happened about 13.8 billion years ago, which we now call the Big Bang (BB). This event occurred with extreme temperatures and as the Universe expanded it cooled. As of today, a sea of photons is observed which is very uniform spatially and has a black body temperature distribution. Matter is also observed as clumped aggregates; suns and galaxies and clusters of galaxies. From these observations, an attempt can be made to extrapolate backwards toward the initial event. Since gravity always "wins" the extrapolation must use a GR formulation.

The Cosmic Microwave Background (CMB) photons have a thermal distribution with a temperature of 2.72 K^o, or an energy of 6.6×10^{-4} eV. There are several basic facts about the Universe which need to be represented. In large enough volumes it is of uniform density and is isotropic.

This conclusion is substantiated by telescopic surveys of millions of galaxies and by the CMB itself. It is also known that all galaxies appear to be moving away from us, the Hubble discovery of Doppler shifts in the spectra of distant galaxies. Since we are not privileged observers, it must be that space itself is expanding, which is possible in GR where the metric is driven by the energy content. These facts allow cosmology to make a "spherical cow" model where the history of the Universe is defined by the time evolution of a single parameter, the scale parameter $a(t)$, which has the dimensions of length, where time is the time on clocks of co-moving observers in free fall. The coordinate system is co-moving and physical distances are set by a scale factor $a(t)$. The metric is isotropic in agreement with data if large enough volumes are smeared out uniformly. This metric is called the Robertson-Walker metric. Under these assumptions the Universe, and it's history is contained in the parameter, $a(t)$.

$$\mathrm{ds}^2 = (\mathrm{cdt})^2 - a^2(t)[d^2(r) + r^2 d^2\Omega] \tag{8.9}$$

Clearly, the metric must be defined by the energy content. Gravity is a very weak force compared to electromagnetism. We can overcome the whole attraction of the Earth with a small bar magnet, for example. A mass that sets the scale when gravity is strong is defined in terms of a "fine structure constant" for gravity. Numerically, the Planck mass, M_P is 1.2×10^{19} GeV. This enormous mass scale indicates when gravity becomes strong. It is not the Schwarzschild radius of a specific mass point but the analogue of the fine structure constant. There is no dimensionless $\alpha = e^2/\hbar c$ for gravity. That means that gravity cannot be theoretically kept under control at all energies and there is no quantum theory of gravity at present.

$$\begin{aligned} &\alpha = e^2/\hbar c \sim 1/137 \\ &\alpha_G = GM_p^2/\hbar c, \quad \alpha_G = 1, \quad M_P = \sqrt{\hbar c/G} \end{aligned} \tag{8.10}$$

Some basic parameters in cosmology are provided in the table below. The current measured value of the Hubble parameter, $H_o = (\mathrm{da}/\mathrm{dt})/a$, with dimensions of inverse time is $2.3 \times 10^{-18}\, s^{-1}$ which defines the Hubble time and the Hubble distance by extrapolating to $t = 0$ assuming that H is a constant, which is not true as will be mentioned, but which gives a rough idea of the spatial and temporal extent of the Universe. The expansion slows as the sources dilute. The CMB temperature is the result of a fit to the Black Body CMB spectrum. The critical density, and the contributions of different types of energy to that density will be explained later.

For now it is important to note that the energy in the Universe has contributions in increasing importance from radiation, ordinary matter, dark matter, and dark energy. The present value of a quantity is denoted by a_o subscript. The Ω values refer to the present. The program is to take current observations and attempt to extrapolate back in time toward the BB. It is to be noted that the model which is required to fit the data contains only about 5% of objects known to the Standard Model, SM, of particle physics.

Selected Cosmological Parameters

Quantity	Present Value
T_o — CMB Temperature (°K)	2.726
$t_H = 1/Ho =$ Hubble Time (sec)	$4.58 \times 10^{17} = 14.53$ Gyr
$D_H = c/Ho =$ Hubble Distance (m)	1.37×10^{26}
ρ_c — critical density $(\mathrm{GeV/m^3})$	5.6
Ω_m — matter fraction (dark matter dominant)	0.315
Ω_b — baryon fraction	0.05
Ω_γ — radiation fraction	4.64×10^{-5}
Ω_Λ — dark energy or vacuum fraction	0.685

The dynamics of the Universe, the R-W metric, is driven by the energy content. The expansion of the Universe is defined by the Hubble parameter H and the density is the total energy density of matter/energy of all types. Note that researchers in this field use cgs units preferentially. Units with $c = 1$ and even $G = 1$ are also often encountered, so the reader of texts should be aware of that. The expression for H assumes the Universe is flat, $\Omega_m + \Omega_\gamma + \Omega_\Lambda = 1$ which will be addressed later. Other contributions to H are, a possible curvature term, $-\kappa c^2/a^2$, and a vacuum energy term $\Lambda^2/3$ respectively.

$$H = (\mathrm{da/dt})/a = \sqrt{(8\pi G\rho + \Lambda c^2)/3} \tag{8.11}$$

The fluid flow equation for the evolution of the energy density in an expanding space is $d\rho/\mathrm{dt}$, showing the loss in density due to the spatial expansion which acts like a cosmic friction term. It is notable that in GR pressure is a source as is mass. With an equation of state, $p/c^2 = \omega\rho$, relating density and pressure, the acceleration of $a(t)$ can be found. For matter $\omega \sim \langle\beta^2\rangle/3 \sim 0$, the Maxwell Boltzmann result, while for radiation, $\omega = 1/3$, $\beta = 1$. The resulting differential equation for $a(t)$ is called the acceleration equation:

$$d\rho/\mathrm{dt} + 3H(\rho + p/c^2) = 0$$
$$\left(\frac{\mathrm{d}^2}{\mathrm{d}t^2}a\right) \Big/ a = -(4\pi G/3)(\rho + 3p/c^2) + \Lambda c^2/3. \tag{8.12}$$

Using the acceleration equation and ρ, p the power law behavior for matter and radiation, shown below, is easily derived, $\rho \sim a^{-3(1+\omega)}$. The energy density scales as inverse a cubed for matter and inverse a to the fourth power for radiation. In both cases the expansion slows the acceleration as the sources become less dense. At large enough values of $a(t)$, matter will become the dominant force over radiation since it dilutes more slowly. The table below lists solutions for pure matter or radiation. In both cases H decreases as $1/t$. The dark energy discussion is deferred for now. The evolution of H is defined by a critical density.

$$\rho_c = 3(H_o)^2/8\pi G, \quad \{H(t)/H_o\}^2 = \rho/\rho_c = \Omega(t). \tag{8.13}$$

Power Law Behavior for Single Sources

	ρ(a)	a(t)	H
Matter	$1/\mathrm{a}^3$	$\mathrm{t}^{2/3}$	(2/3)/t
Radiation	$1/\mathrm{a}^4$	$\mathrm{t}^{1/2}$	(1/2)/t
Dark Energy	constant	e^{Ht}	constant

The following code plots the density and temperature for radiation and matter using current cosmological data. The CMB provides the present photon density. The mass density is taken to be both mass and dark matter, which is required by data on the rotation curves of galaxies and by observation of astronomical mass lensing. They require a matter source in excess of what is directly measured by counting up stars and estimating their mass given their luminosities. The time when the densities are equal is illustrated, about 24,000 years since the Big Bang, which is taken to be $t = 0$, where $a(0) = 0$. At that time the radiation temperature was about 18,000 K^o, with an energy $\sim 1.5\,\mathrm{eV}$. The number density of photons in the CMB at present is about $4.1 \times 10^8 \gamma/m^3$. There is an assumed abrupt shift in the behavior of the expansion when the densities are equal. Later work will use MATLAB utilities to make the true smoother behavior evident. At $t \sim$ a few seconds, the radiation temperature is on a scale of nuclear binding energies, which means that nuclear physics will be needed to extrapolate backward to earlier times.

```
% find flat space quantities in matter and radiation dominated eras
% ignore dark energy, DE, effects, important at late times
% get present value of H accurately known ?
% present Ho in km/sec*million lyr (~30);
% h (~0.73), to ~ 13.8 Gyr;
% assume omega = 1 (flat space)
% To = CBR temp at present
% Ho = 30 .*h , LH = c/H, tH = 1/H, t = n/H, vH = c(1+q),
q = (1-n)/n,
% present time cto
n = 2.0 ./3.0 ; % matter dominated, flat GR, a dependence on t,
a/ao=(t/to)^n
H = 30 .*0.73;
h = 0.73; % present Hubble constant
To = 2.73; % CBR temp in deg K
% Hubble Time in byr ;
tH = 300 ./H  % Hubble Time in byr
```

```
tH = 13.6986
```

```
Teq = To .*(tH ./teq) .^0.6666;   % temp of CBR at equal density
% Energy at teq (eV) %
Teq./(40 .*300)
```

```
ans = 1.5463
```

```
% Temperature at teq - degree K ;
Teq
```

```
Teq = 1.8555e+04
```

```
% c = 3 x 10^5 km/sec, 1 yr = 3.16 x 10^7 sec,
% 1ly = 9.46 x 10^12 km
% convert to sec, sec-1
%  Time at Equal Density (no v no DE) - sec ;
teqs
```

```
teqs = 7.6907e+11
```

```
% extrapolate back in rad dominated era, matter n = 2/3,
rad n = 1/2
% Matter Density at That Time
% Radiation Density at That Time,rhore
```

```
% Radiation Temperature at That Time ,Tearly
% now plots of interesting quantities
```

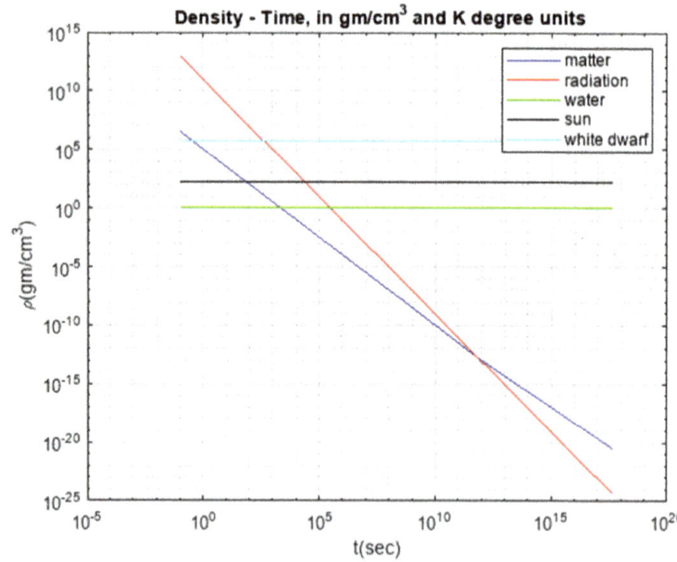

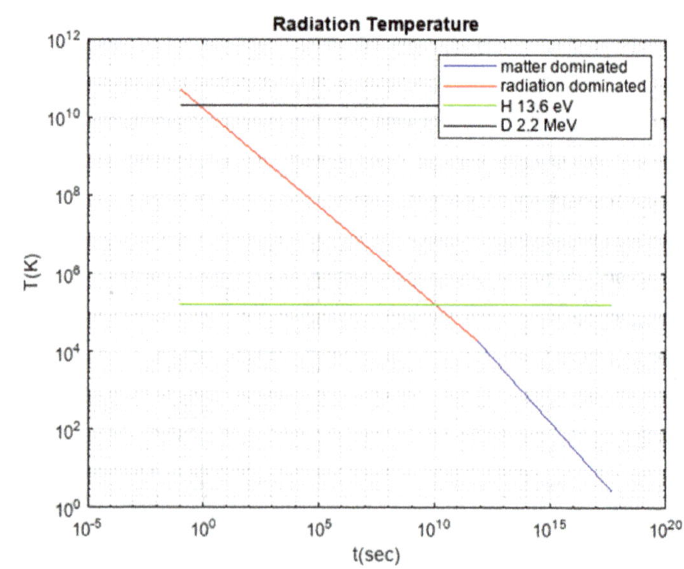

Figure 8.7: Top — Density of radiation and matter as a function of time. Bottom — radiation temperature as a function of time.

8.6 Smooth Radiation/Matter Transition

The Hubble parameter is defined to be $H(t) = (\mathrm{d}a(t)/\mathrm{d}t)/a(t)$. Scaling to the present time, $t \to t_o$, $H_o = H(t_o)$, $\alpha(t) = a(t)/a_o$. The critical density for a "flat" Universe, with zero curvature is $\rho_c = 3H_o^2/8\pi G$. The value of H at other times is shown below. The mix of contributors to $H(t)$ will change with time, but a flat space will remain flat, $\rho = \rho_c$, and it is assumed, in conformance with the data, that the Universe is flat, for some reason. Ignoring curvature, assumed to be zero, the general behavior of the scale factor $\alpha(t)$ is:

$$(d\alpha/\mathrm{d}t)/a = H_o\sqrt{\Omega_\gamma/a^4 + \Omega_m/a^3 + \Omega_\Lambda} \tag{8.14}$$

The Ω values are the present ones, as is the H_o value and the scale at present is a_o. Alternatively, one can solve for $H(t)$ instead of $\alpha(t)$ by using the definition $H(t) = (d\alpha/\mathrm{d}t)/\alpha$. These expressions can be used to solve for the scale factor as a function of time in the presence of several sources.

MATLAB has the tools to easily solve for 2 combined sources of expansion, in this first case both matter and radiation. Now a small correction for the primordial neutrinos is also made. Symbolic definite integration is employed to find the Hubble parameter implicitly. Explicit solutions are possible in terms of the square root of the strength of the source and $H_o t$ raised to a power for radiation and matter and exponentiated in the case of vacuum energy. The power law scaling is reproduced, but here with specific constants. The expressions can also be easily inverted. Note that t is always less than the Hubble time due to deceleration of the expansion.

$$\begin{aligned} \alpha_\gamma &= [(3/2)\sqrt{\Omega_m}H_o t]^{2/3}, & H_o t &= \alpha_m^{3/2}(2/3)/\sqrt{\Omega_m} \\ \alpha_m &= [2\sqrt{\Omega_\gamma}H_o t]^{1/2}, & H_o t &= \alpha_r^2/2\sqrt{\Omega_\gamma} \\ \alpha_\Lambda &= \exp(\sqrt{\Omega_\Lambda}H_o t), & H_o t &= \ln(\alpha_\Lambda)/\sqrt{\Omega_\Lambda} \end{aligned} \tag{8.15}$$

With more than a single source an explicit calculation can be made or an implicit approach is also possible. Formally, the expression for $d\alpha/\mathrm{d}t$ is integrated. The calculation for matter and radiation as a function of α as an explicit calculation is displayed. The transition from RD to MD is not abrupt but extends over 3 orders of magnitude of the cosmological scale factor α. The transition from $Ht = n$ is from $n = 1/2$ to $n = 2/3$. The vertical red line shows when the 2 effects are equal which is where the abrupt transition was assumed previously. The results are:

$$\begin{aligned} &H(t) = H_o\sqrt{\Omega_m/\alpha^3 + \Omega_\gamma/\alpha^4} \\ &H_o t = (1/3\Omega_m^2)\{4\Omega_\gamma^{3/2} - 6\Omega_\gamma\sqrt{\Omega_\gamma + \Omega_m\alpha} + 2(\Omega_\gamma + \Omega_m\alpha)^{3/2}\} \end{aligned} \tag{8.16}$$

```
% look at flat space with a mixture of matter and radiation energy
% omegm = 0.315, omg = 4.65e-5 x 1.68 (CMB+3 generations of v, all
relativistic energy)
y = int(x/sqrt(a*x+b),x); % y = Hot, a = omegm, b = omegg
% definite integral fron alpha = 0 to alpha
z = -(4*b^(3/2))/(3*a^2);    % lower limit of integral
yy = y - z;
Hot = yy;
Hot = subs(Hot,a,Om_m); Hot = subs(Hot,b,Om_g); Hot = subs(Hot,x,al)
```

$$\text{Hot} = \frac{2(\text{Om}_g + \text{Om}_m\text{al})^{3/2} - 6\text{Om}_g\sqrt{\text{Om}_g + \text{Om}_m\text{al}}}{3\text{Om}_m^2} + \frac{4\text{Om}_g^{3/2}}{3\text{Om}_m^2}$$

```
% Symbolic implicit Integral for Hot, x = alpha, a = omega m, b =
omega gam
a = 0.315;
b = 4.65e-5 .* 1.68; % CMB radiation + 3 v generations
Hott = eval(yy);
Ht = Hott.*sqrt(a ./(all .^3) + b ./(all .^4));
% plot HT as the variable ~ n in power law
```

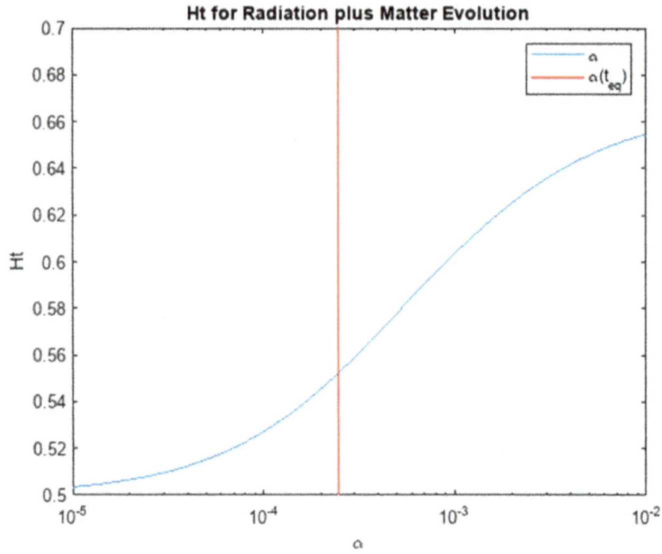

Figure 8.8: *Ht* as a function of the scaled variable $a(t)/ao$, showing a smooth transition from *Ht* of 1/2 to *Ht* of 3/2.

8.7 Solar Nucleosynthesis

Early times are clearly expected to be radiation dominated. At scales of the Planck mass, the physics is unknown. The earliest times that are understood by particle physicists are at the TeV energy scale. There would be a photon sea with a small admixture of quarks, leptons and W, Z and Higgs weak bosons. As cooling progresses the quarks bind together to form the lowest mass protons and neutrons. The lightest leptons, the electrons remain along with the weakly interacting neutrinos. Known physics should be able to predict the behavior of the Universe from this mass scale down to the present and to extrapolate to predict the future.

Having successfully modelled the Sun, one can assume that the physics of stars is reasonably well known in the equilibrium state. However, it is necessary to understand the detailed composition of different nuclei in the Sun. In order to do that, it is necessary to understand the initial creation of the nuclei in the Sun. The basic assumption is that there was a "Big Bang", BB, which was initially a fireball of photons with a small admixture of matter, electrons, neutrinos, and nucleons. The temperature of the fireball was, at short times after the BB, so high that only lowest mass particles existed. Any more complex bound states of these objects would be broken apart by the bath of high energy photons.

For now define a time zero to be the BB time. The Universe expands as a fireball with a parameter H, the Hubble parameter, with dimensions of inverse time, such that $1/H$ is $\sim$ the age of the Universe. Taking the current measured value of H, H_o the time is $\sim$14 billion years. The fireball cooled in the expansion until more complex objects could exist without being continuously broken apart by the energetic photons. The strongest force, the nuclear force, has the largest binding energies. Because of that fact, the start of the history of the Universe begins with the formation of nuclei. This process happened within "the first 3 minutes".

The photons from the Big Bang are initially by far the dominant particle in the fireball and they expand and cool. They are in thermal equilibrium with the trace matter, neutrinos, electrons and nucleons. A light nucleus has a typical binding energy of about 8 MeV per nucleon. There are so many photons that nuclei cannot form above a photon temperature of about 1 MeV since they would be broken up by the photons in the high energy tail of the Planck energy distribution. For a temperature $T \sim 1$ MeV only nucleons are in thermal equilibrium with the photons. At that temperature the neutrinos decouple and propagate independently. Detailed estimates of the freeze-out of neutrinos yield an estimate of 3 neutrino generations, in agreement with the current Standard Model of high energy particle physics.

For the nucleons a precise calculation can be made because the masses and cross sections are all well measured in the laboratory. At a temperature of ~0.8 MeV the neutrons and protons decouple from the photons, with a relative abundance given by their mass difference, $n_p/n_n = e^{Q/\mathrm{kT}}$, $Q = m_n - m_p = 1.293\,\mathrm{MeV}$. The neutrons then decay and are also taken up into the formation of nuclei. The mass fraction of a species, X, is the abundance relative to all nucleons, which for neutrons is: $X_n = n_n/(n_n + n_p)$. The Boltzmann transport equation for the neutrons is driven by the n-p reaction rate, Γ_{np}, $p + \nu + e^- \longleftrightarrow n$;

$$\mathrm{d}X_n/\mathrm{dt} = \Gamma_{\mathrm{np}}[(1 - X_n)e^{-Q/\mathrm{kT}} - X_n] \tag{8.17}$$

The n-p reaction is a weak interaction process, which scales as T^5. At the n freeze-out temperature of ~0.8 MeV the n/p ratio is ~0.2. The freeze-out abundance of n is about $n_p/n_n = 5$. After this time the n decay and the n fraction at the time of Helium formation is only ~11% when T is ~0.3 MeV. Since there are $2n$ in each He nucleus, an estimate of the primordial He abundance with respect to all nucleons is ~22%. A slightly more detailed estimate of the primordial Helium abundance can be made. A simple model, nuclear statistical equilibrium, is used to estimate the abundance of light nuclei. The nucleon to photon ratio is taken to be $\eta = 5.5 \times 10^{-10}$ from knowledge of the present Cosmic Microwave Background (CMB). There is a competition in nuclear formation between the binding energy B_A and the photons to form a nucleus with Z protons and A-Z neutrons. The small value of η (photon entropy) means that nuclei freeze out at temperatures well below their binding energies. At a given temperature, T, Z protons and (A-Z) neutrons come together and bind with an energy B_A in the presence of a large sea of photons which may break the nucleus up again.

$$X_A \sim T^{3(A-1)/2}\eta^{(A-1)}X_p^Z X_n^{(A-Z)}e^{B_A/T}. \tag{8.18}$$

Specifically n and p form deuterium D. The abundance of D is small because it's binding energy is only 2.2 MeV which makes it very sensitive to η, $n + p \longleftrightarrow D + \gamma$. The protons are largely taken up in He formation as are the neutrons. The simplified set of approximate Boltzmann like equations to be solved for are.

$$\begin{aligned}
&X_A = n_A A/n_N, \quad 1 = X_p + X_n + X_D + X_{\text{He}} \\
&X_n = X_p e^{-Q/T}, \\
&X_D \sim T^{3/2}\eta e^{(B_D/T)} X_p X_n \\
&X_{\text{He}} \sim T^{9/2}\eta^3 e^{(B_{\text{He}}/T)} X_p^2 X_n^2
\end{aligned} \tag{8.19}$$

The n follow these equations and do not freeze out in this approximation, nor do they decay. The p become dominant until taken up by He formation, which abundance rises very rapidly with falling T to about 22% at $T \sim 0.28\,\text{MeV}$. The script simply assumes 1/2 n and 1/2 p at high temperatures and then evolves to lower T.

```
% evaluate mass fractions in Nuclear Statistical Equilibrium
% Follows Kolb - Turner text
% Light Elements in NSE
% starting values, only p and n at high T
y1(1) = 0.5; % p
y2(1) = 0.5; % n
y3(1) = 0.0;  % D2
y4(1) = 0.0;  % He
Q = 1.293; % n-p mass difference
B2 = 2.22 ; % binding energy of D, He
B4 = 28.3 ;
mn = 938.0; % nucleon mass, MeV
% entropy - nucleon to photon ratio, photons from from CMB + v
eta = 7.15e-10;
% a simple numerical rundown of the system of equations
```

The plots are not correct in detail. However, they show p take-up to feed He production and the rapid rise of Helium at T scales much less than the binding energies because of the large photon bath in which the nucleons are immersed. $T \sim 0.2\,\text{MeV}$. They also show the very small D abundance and the large He abundance due to the large He binding energy.

Instead of the 22% He abundance expected the Sun has 28%. Indeed the Sun is not a first generation star. Such a star would be predicted to be mostly protons with 22% Helium and very little Deuterium. Heavier elements are expected to be blocked by the large He binding energy. Indeed, heavier elements up to iron are thought to arise in older stars that have used up the light elements to fuel the fusion reactions, or as supernova remnants

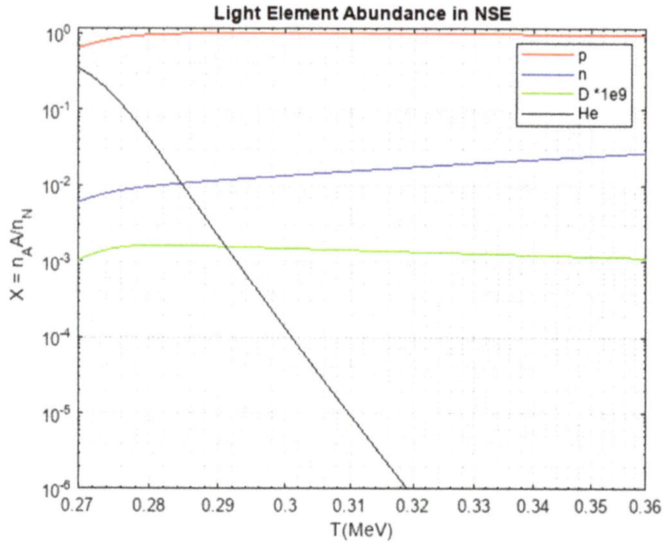

Figure 8.9: The abundance of light elements as a function of the radiation temperature.

which are accreted into second and later generation stars. Elements heavier than iron can only arise from supernovae because iron has the peak binding energy per nucleon and energy cannot be liberated by the fusion of iron.

Using the data from nuclear physics, an extrapolation from the present to the CMB formation will next be explored at the eV energy scale and the solar nucleosynthesis at the MeV scale is successful. At higher energy scales, laboratory data at the TeV scales is available. That data is codified in the Standard Model of particle physics. That model is now complete, with no indication of new physics at higher energies. Indeed, higher energies are not yet available in the laboratory, and the scale of the Planck mass is clearly out of reach. That being the case a closer look at the available data seems to be called for.

8.8 CMB, the Saha Equation, and Ω_b

Previously the Helium fraction was estimated in the discussion of nucleosynthesis, which occurred at energies of $\sim$1 MeV, less than the Helium binding energy due to the large photon entropy. The fluids then continued to evolve, as a mix of the dominant photons, with minority protons and Helium nuclei along with electrons and weakly interacting neutrinos. A similar situation occurs as further expansion cools the photons and finally stable atoms with binding energies ~13.6 eV can form. In this case when the plasma scattering rate is, $\Gamma_\gamma < H$, the photons fall out of equilibrium and approximately free stream from to the present time. The photon fluid is transparent and the CMB appears at a mean photon energy ~0.26 eV, much less than the Hydrogen binding energy, at a time of ~400,000 years. The CMB observed today is the earliest time that can be observed using light since the plasma was previously optically opaque. If primordial neutrinos could be observed, their weak interaction scale of ~MeV would open a view to times of ~ minutes. With the advent of gravitational wave detection, the possibility of "seeing" much earlier than the CMB decoupling time becomes possible. However, the large wavelengths make the construction of "antennas" more difficult.

As with the case of nucleosynthesis, a simplification can be made, and the full Boltzmann transport equations are not solved here, $e + p \longleftrightarrow H + \gamma$. The relevant cross section is that for Thomson scattering, $\sigma_T = (8\pi/3)\alpha^2\lambda_e^2$. The relevant binding energy is $B_H = 13.6\,\text{eV}$ for atomic Hydrogen. The simplification is to assume thermal equilibrium and use the Maxwell-Boltzmann distribution. The number density ratio of e, p and H are then:

$$n_H/(n_e n_p) = [(m_e T)/2\pi]^{-3/2} e^{B_H/\text{kT}} \tag{8.20}$$

Define X_e to be the fraction of free electrons, $n_e/(n_e + n_H)$. The binding energy competes with the photon entropy $n_b/n_\gamma = n_p/(X_e n_\gamma)$. Solving for X_e results in a quadratic equation, the Saha equation:

$$1 - X_e = X_e^2\left[4\sqrt{2/\pi}\zeta(3)(\eta)(T/m_e)^{3/2} e^{B_H/\text{kT}}\right] \tag{8.21}$$

The decoupling of the plasma occurs a $\text{kT} \ll B$ as was the nucleosynthesis case. The scale factor used in the script is $1 + z = 1/\alpha(t) = a_o/a(t)$ but the results are displayed as $a(t)$. The CMB decoupling occurred when the Universe was ~1400 times smaller than at present. As shown in the plot below, the decoupling time is a function of the baryon fraction of the critical density, which serves as crucial input to a consistency check of the present cosmological model. The user can choose the Ω_b value in order to see the effect on the decoupling time.

```
% look at CMB decoupling in the Saha approximation to the complete
Boltzmann eq
kb = (1.0 ./40) .*(1.0 ./300);  % Boltzmann k in eV/oK
h = 0.73;  % Hubble factor(3)
```

```
To = 2.75;    % present temp in Kelvin
B = 13.6 ;   % H binding energy (eV)
me = 511000.0 ; % electron mass in eV
zet = zeta(3);  % Riemann function
%
omegab = 0.05 ; % Present Baryon Density/Critical = Omega
% span of z around CMB decoupling, 1+z = a(t?/ao
eta = omegab .* h .*h .*2.7 .*10 .^-8; % baryon to photon ratio
y = (4.0 .*sqrt(2) .*zet) ./sqrt(pi);
y = y .*eta .*exp(B ./(kb .*T));
y = y .*(((kb .*T) ./me) .^1.5);
q = 1.0 ./y;
Xe = (-q + sqrt(q .*q + 4.0 .*q))./2.0; % quadratic Saha solution
% find 50% point, define as decoupling
```

```
zdec = 1.4141e+03
aldec = 7.0664e-04
```

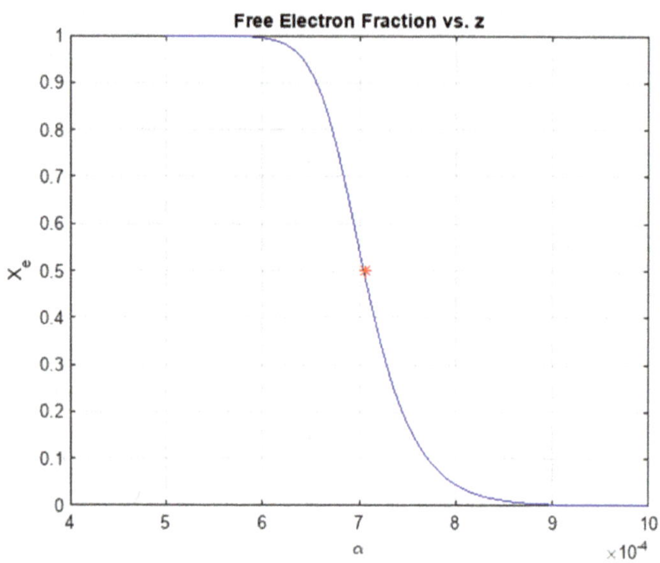

Figure 8.10: Electron abundance as a function of the scaled variable $a(t)/ao$. The red star shows the 50% point for electrons.

```
% look at decoupling dependence on Omega b
etaa = omegbb .* h .*h .*2.7 .*10 .^-8; % baryon to photon ratio
```

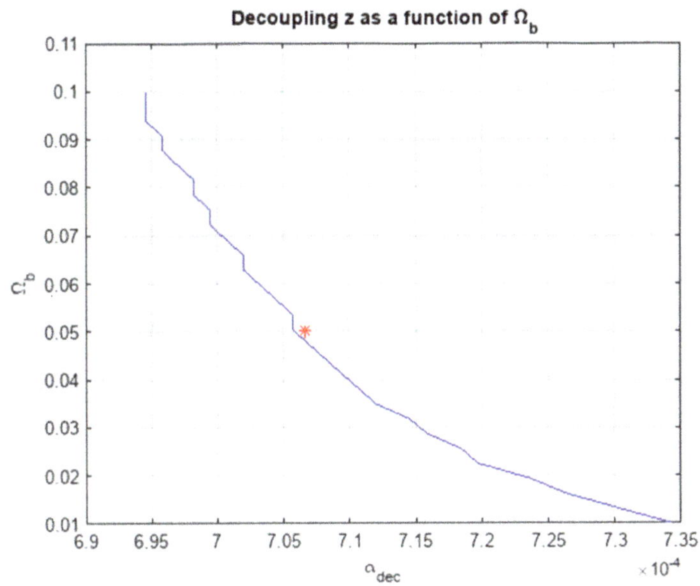

Figure 8.11: Baryon (p) critical fraction as a function of the scaled variable $a(t)/ao$. The red star shows the accepted value derived from all cosmological data.

8.9 Cosmic Microwave Background (CMB)

This Section has moved from planetary voyages, to models of stars and their elemental abundances, and finally to the Universe. This progression is data driven at each step. Data on the photon temperature in different locations over the entire sky are shown below. It is not to say that the temperature is very variable. The fluctuations across the sky are in parts per 10 million. In fact the CMB is almost uniform over the entire sky.

The data are represented in a more compact form as the angle-angle intensity correlations, also shown below. The most prominent structure is a peak at a multipole moment value of l of about 240 which is an angular scale less than, but comparable to 0.4 degree. There are also higher order acoustic peaks at approximately integer multiples of the first peak, higher harmonics.

The basic facts about the CMB are the uniformity followed by the angular correlations and they need to be explained when making a model of the data. The uniformity is a mystery since causally disconnected parts of

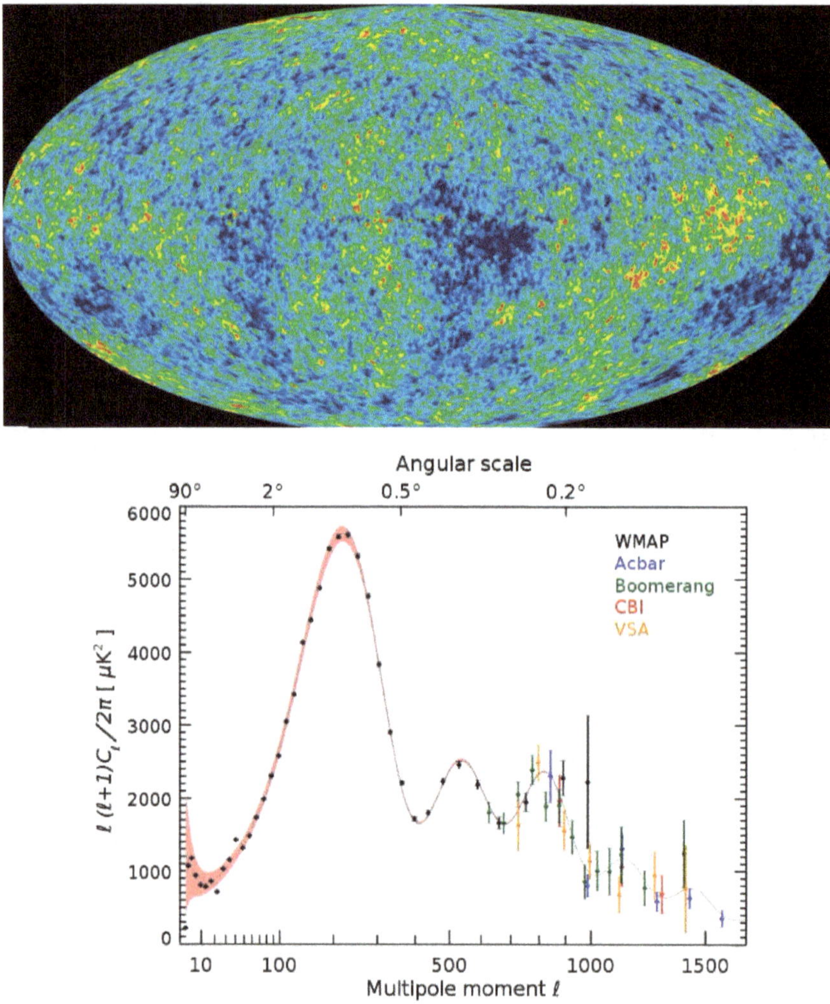

Figure 8.12: Top — Data on the temperature variations of the CMB over the entire sky on a scale of parts per million. Bottom — Strength of the multipole moments of the angular CMB temperature variations.

the sky have the same temperatures. This fact has led to the hypothesis of "inflation" where some mechanism caused the very early Universe to expand more rapidly than the speed of light which allows for the CMB uniformity to be understood. This hypothesis is an entire subject which will not be covered in this text but must be noted. It is clear, however, that the specific mechanism for such "inflation" is not yet understood.

The CMB is described by making a model using, at the simplest level of approximation, a set of 5 Boltzmann transport equations. Classically, a Boltzmann equation for the time development of a fluid of density ρ with a velocity $\vec{v}$ which has a scattering process, σ, with some constituents is: $\frac{\partial}{\partial t}\rho + \vec{v} * \vec{\nabla}\rho = \int(d\sigma/d\Omega)d\Omega$. A simplified Boltzmann equation was previously invoked in the exploration of the formation of Helium and other light nuclei. The densities in the specific case of the CMB refer to the components of the Universe needed in order to get a good fit; dark matter, the baryons, the photons, electrons, neutrinos and the dark energy. At the times in question for the CMB the dark energy plays little part in the CMB evolution. There is a mix of 2 basic fluids, the dark matter and a photon-baryon plasma. The baryons are tightly coupled to the dominant photons. There is a competition between the photon radiation pressure and the baryon gravitational compression. That tension leads to sound waves or acoustic oscillations of the medium. Those oscillations are frozen at decoupling, when the interaction rate of the baryon-photon plasma becomes less than H. The speed of sound in the plasma is $c_s/c = \sqrt{(1+\Omega_b)/3}$, which depends on the baryon content. The dark matter does not interact with the photons, being "dark".

The matter, protons and electrons, and photons become of equal importance as sources for the Hubble parameter, H, about 24,000 yr after the Big Bang (BB) at a temperature of about 1.6 eV. The plasma expands and cools. The decoupling is due to the expansion, when $\Gamma_\gamma < H$. The dilution caused by expansion means that thermal equilibrium is lost. The photon temperature is about 0.26 eV when that occurs, about 400,000 years after the BB. After this the photons propagate freely, imprinted by the acoustic oscillations at decoupling. This is then the earliest time available to optical probes of the Universe, such as telescopes, since before that the photons are strongly coupled to the baryons. Numerically, the fundamental acoustic peak $\sim$ fills the Universe at decoupling so that $l \sim 1/\alpha(\text{CMB}) \sim 1000$ with harmonics at higher values. With a more careful, but still simplified, analysis, seen below, the peak at a multipole value of $l \sim 240$ is predicted.

There is sufficient information in the CMB shape to determine each of the cosmic parameters with no degeneracy and a good fit is obtainable. The user can vary 2 of the basic parameters here, the matter and the baryon densities, and see that each combination gives a unique shape and size to the first 2 acoustic peaks seen in the multipole moment plot. The details are not too important and the model is simplified. The Boltzmann equations are solved using "ode45" and the numerical integration utility "trapz". What is important is that the CMB, in detail, is reproduced with only 5 distinct contributions, at least in this simplified model. Much more complex models exist, but the basic components are usually limited to ordinary matter, radiation, dark matter and dark energy. It is amazing that the complexities of the CMB can be modelled by such a limited set of basic constituents of the Universe.

```
% constants for the system of 5 equations
% 2 for CDM, 2 for photon-b fluid, 1 for the potential and deriv
omm =0.25% dark matter, oml = 0.69 - dark energy fixed
```

```
omm = 0.2500
```

```
omb =  0.05    % baryons
```

```
omb = 0.0500
```

```
adec = 1.0 ./1100;    % recombination/decoupling scale a
% SM a(t) at equality  decoupling and alpha parameter ,aeq, adec
% omega rel fixed, compute aeq for this input, assume adec is fixed
% This a(t) at equality, decoupling and alpha parameter,aeq,
adec, al
% conformal time at present, decoupling and RD_MD equality,tauo,
taudec, taueq
% time ratio present/decoupling tauo ./taudec);
tauoo = 35; % project to present - 49 with 3 v and SM
% conformal time ratio of present to dec,tauoo
% approximate damping scale in kappa
% t is the conformal time scaled to the conformal time at dec (x )
% project Fourier onto Ylm with spherical Bessel functions
% integrate over k to get Cll
pol = polyfit(ll,Cll,11);
Ctt = polyval(pol,ll);
% first acoustic peak - height and location
% First Acoustic Peak Height at l = cpeak, lpeak
```

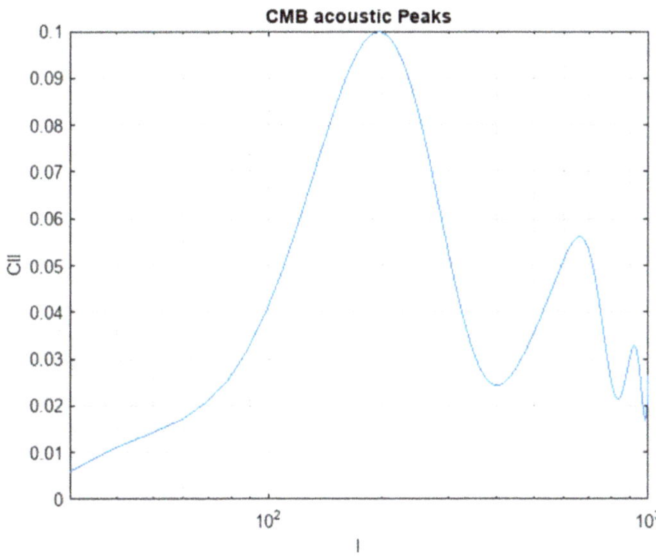

Figure 8.13: Correlation in angle-angle as a function of l for a specific choice of the baryon fraction. The first 2 acoustic peaks are evident.

```
cpeak
```

```
cpeak = 0.0999
```

```
lpeak
```

```
lpeak = 196.5657
```

8.10 Matter and Vacuum Energy

Observations, such as galaxy rotation curves, or gravitational lensing, have driven model builders of the Universe to postulate the existence of "dark matter", which gravitates but otherwise is not yet proven to otherwise interact. Dark matter is dark; it does not have strong or electromagnetic interactions. The laboratory searches assume the dark matter has weak interactions with ordinary matter and can be detected in laboratory experiments. It is notable that the Standard Model of high energy particle physics has no candidate for dark matter.

More recently, supernova surveys, using the supernovae as "standard candles" have shown that the expansion of the Universe has, quite recently, begun to speed up. Since known matter and energy slows the expansion, a new "dark energy" has been postulated which acts gravitationally but repulsively. It's identity is also unknown, Cosmologists now have a successful model used to describe the Universe in detail, but one which postulates 2 new entities, comprising ~95% of the energy of the Universe but which have no known identities among the members of the Standard Model of fundamental particles and fields.

The metric could also have curvature, with the term d^2r modified to b, $d^2r/(1-\kappa r^2)$. Experimentally, the curvature is zero within measurement error, which means the Universe is flat and the sum of all $\Omega = 1$. Since the sum of all matter, normal plus dark, is too small, the dominant contribution to the Ω sum is ascribed to dark energy. The simplest assumption, so far confirmed by experiment within errors, is that this contributor appears in the Einstein equations as a $\Lambda g_{\mu v}$ term, a vacuum energy, or a "cosmological constant" which does not evolve. Vacuum energy also has no known laboratory candidate. However, since dark energy must have the quantum numbers of the vacuum, it is notable that the recently discovered Higgs boson has those quantum numbers. It is, in fact, the only known scaler particle contained within the Standard Model of fundamental particles. So fundamental scalers exist and perhaps the dark energy is composed of such particles.

If this is a correct model, instead of looking back in time to the Big Bang about 13.8 billion years ago, one can also look forward and predict the future evolution of the Universe. As the other forms of energy all dilute with expansion, the vacuum energy being the metric itself, will ultimately dominate. The vacuum energy or dark energy by itself has a scale factor which diverges, $\frac{\mathrm{d}^2}{\mathrm{d}t^2}a = a(\Lambda/3)$, $a(t) \sim e^{(\sqrt{\Lambda/3})t}$. The dark energy is observed to be repulsive, although it need not be and was introduced by Einstein originally to be attractive and to make the Universe static. The full dynamical equation for $\alpha(t)$ was previously shown. The vacuum energy fraction Ω_Λ is chosen by the user. By itself the vacuum energy has a solution, $H_o t = \ln(\alpha_\Lambda)/\sqrt{\Omega_\Lambda}$ The equation for $H_o t$ in the presence of both matter and dark energy also has a closed form solution.

$$
\begin{aligned}
&H(t) = H_o\sqrt{\Omega_m/\alpha^3 + \Omega_\Lambda}, \quad \alpha = a(t)/a_o \\
&H_o t = (2/3\sqrt{\Omega_\Lambda})\ln[(\sqrt{\Omega_m + \Omega_\Lambda\alpha^3} + \sqrt{\Omega_\Lambda\alpha^3})/\sqrt{\Omega_m}]
\end{aligned}
\tag{8.22}
$$

It appears that if the model is correct we are at present in an accelerating Universe driven by vacuum energy. The green dot in the plot below is the present. If a repulsive cosmological term exists it will always win in the long run as the matter and radiation importance are steadily diluted by the expansion of space.

```
% now look at flat space with a mixture of matter and vacuum energy
% supernova + CMB wiggles indicate omegav ~ 0.685, omedm = 0.315
Ho = 30; % present Ho in km/sec*million lyr
h = 0.69; % estimate for h
omegv = 0.7; % Value for Vacuum Omega = (0,1)
% Ho in km/sec*million ly, Hubble length and time and critical
density
% Ho = 30 .*h , DH = c/H, tH = 1/H,
Ho = Ho .*h; % present Hubble constant
tHo = 300 ./Ho  % Hubble Time, in byr
```

```
tHo = 14.4928
```

```
% specify the fraction of critical density in the vacuum
% present time t, not n/H due to vacuum energy
% pick a/ao = alpha and solve for t
% compare to no vacuum energy
```

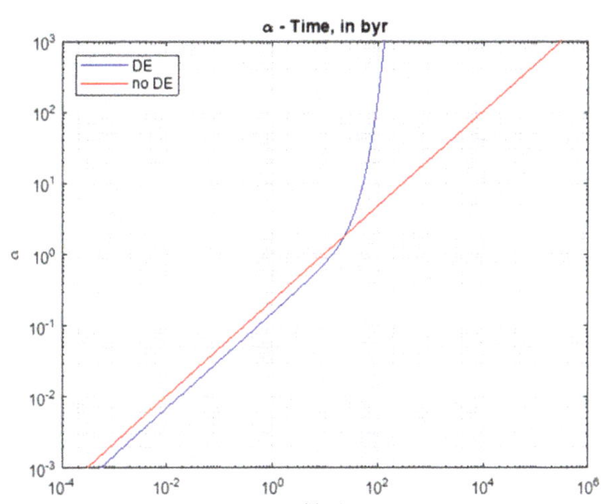

Figure 8.14: Dependence of $a(t)/ao$ as a function of time (top) and $H(t)/Ho$ as a function of $a(t)/ao$ (bottom) for a combination of dark energy and matter (blue) or matter alone (red). A smooth transition from power law with $-2/3$ exponent to dark energy dominance occurs with a crossover indicated with the green dot.

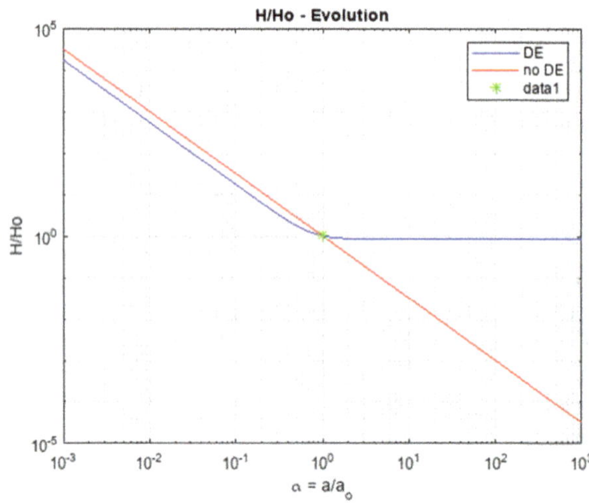

Figure 8.14: (*Continued*)

8.11 The Higgs Boson

The major particle physics discovery since this book was first published more than a decade ago is the discovery of the Higgs boson. This fundamental particle is a scaler. It has the quantum numbers of the vacuum. It is the sole such particle of all the fundamental particles of the Standard Model, SM, of particle physics. This unique particle would have cosmological implications because it is an intrinsic part of the vacuum. Therefore it needs to be explored in a bit more depth. For a free scaler field ϕ the Lagrange density is $l = \phi(p_\mu p^\mu - (\mathrm{mc})^2)\phi$, where p_μ is the 4 vector momentum and m is the mass. The basic form is familiar from the relationship, $E^2 = (\mathrm{cp})^2 + (\mathrm{mc}^2)^2$ with a SR vector with components $p_\mu = (E, \overrightarrow{\mathrm{cp}})$. The Higgs boson has self interactions which are defined to have quadratic and quartic dependence on the Higgs fields. $V(\phi) = \mu^2\phi^2 + \lambda\phi^4$, parameterized as a Taylor expansion in the fields. This potential has a minimum at a field value $\langle\phi\rangle^2 = -\mu^2/2\lambda$ and the potential minimum is;

$$V(\langle\phi\rangle) = \mu^2\langle\phi\rangle^2 + \lambda\langle\phi\rangle^4, \quad V_{\min} - \mu^4/4\lambda. \tag{8.23}$$

The physical ground state Higgs boson appears when one expands about the potential minimum, which is the minimum of the true vacuum,

$\phi = \langle\phi\rangle + \phi_H$. The potential is then;

$$V(\phi) = -\lambda\langle\phi\rangle^4 + 4\lambda\langle\phi\rangle^2\phi_H^2 + 4\lambda\langle\phi\rangle\phi_H^3 + \lambda\phi_H^4. \tag{8.24}$$

These terms can be identified from left to right as the constant potential minimum, a Higgs mass term with mass squared $4\lambda\langle\phi\rangle^2$ and the Higgs self-interactions involving 3 and 4 Higgs at an interaction vertex - triplet and quartic couplings with coupling constants λ. Indeed these predicted couplings are being very actively searched for at colliders such as the Large Hadron Collider (LHC) at CERN. Such couplings for the electroweak force carriers, the W and Z bosons, have been observed already at the LHC, for example with the observation of W pair production indicating triple couplings. Evidence for triple W production, also has indicated the existence of quartic couplings. These predictions of the Standard Model of particle physics and their observation is a confirmation of the SM. Presently there is a program to search for the production of pairs of Higgs, to see if the predicted strength of the triplet coupling is confirmed.

For the potential itself, it is defined by 2 parameters, μ, λ. One parameter is measured by using the boson masses, γ, g, W, Z where the photons and gluons are required to be massless. For the weak force carriers, W and Z, their mass is supplied by the Higgs since they couple to $\langle\phi\rangle$. The second parameter, λ is measured once the Higgs particle mass is determined, which completes the determination of the Higgs potential, assuming no higher order terms exist.

```
syms V u l ph ph2 pH
V = u^2*ph2 +l*(ph2^2) % potential
```

V = $l\text{ph}_2^2 + \text{ph}_2 u^2$

```
eqn = diff(V,ph2) == 0; % minimize
sol = solve(eqn,ph2) % phi^2 for min V
```

sol = $-\frac{u^2}{2l}$

```
Vm = subs(V,ph2,sol) % V at min
```

Vm = $-\frac{u^4}{4l}$

Using the assumed potential, a minimum occurs at $\langle\phi\rangle^2 = -\mu^2/2\lambda$, $V_{\min} = -\mu^4/4\lambda$. Having found the minimum, expand around it with the physical Higgs field ϕ_H, $\phi = \langle\phi\rangle + \phi_H$. The result for $V = \mu^2[\langle\phi\rangle + \phi_H]^2 + \lambda(\langle\phi\rangle + \phi_H)^4$, which contains a mass term for the Higgs, quadratic in the Higgs field, with mass $M_H = 2\sqrt{2\lambda}\langle\phi\rangle$. The Higgs also induces a mass in the weak gauge bosons. The measured mass of those weak bosons determines the parameter, $\langle\phi\rangle$ to be 174 GeV. The remaining parameter is found with the discovery of the Higgs boson itself, at a mass of 125 GeV, which determines $\lambda = 0.50$. That is not to say that major questions like dark matter and dark energy are not yet to be understood. [parenthetically the Standard Model has symmetry group structures, SU(3) × SU(2) × U(1) and the present treatment has ignored that fact, which necessitates a tweak for the expression for the Higgs mass because of the group "Clebsch-Gordan" factors]

```
% phiminsq= -mu*mu/(2*lambda)
% numerical constants
% phimin = 174 GeV, lam = - 0.26 (tweaked)
phimin = 174;
lamda = 0.88 ;
mu = -sqrt(2 .* lamda .*(phimin .^2));
Vmin = -(mu .^4) ./(4 .*lamda)
```

```
Vmin = -8.0664e+08
```

```
phi = linspace(0, 300);
V = -mu.*mu.*phi.*phi+lamda .*phi.*phi.*phi .*phi;
plot(phi, V,'-b')
```

The Higgs has a minimum nonzero field, $\langle\phi\rangle$ of 174 GeV. This is a unique attribute in particle physics and induces a mass to the Higgs and indeed to all the other fundamental particles. The Higgs vacuum expectation value and the coupling of the Higgs to all other fundamental particles is responsible for the mass of those particles. The actual Higgs field, ϕ_H as an elementary particle has attributes that become clear when the field is expanded about the minimum, $\langle\phi\rangle$ about which the field excitation oscillates, $\phi = \langle\phi\rangle + \phi_H$. Terms quartic, triplet, and quadratic in ϕ_H are evident. The mass of the physical Higgs particle, oscillating about the minimum of the potential, is 125 GeV. The red line is added simply to schematically indicate the Higgs boson oscillating about the minimum of the potential.

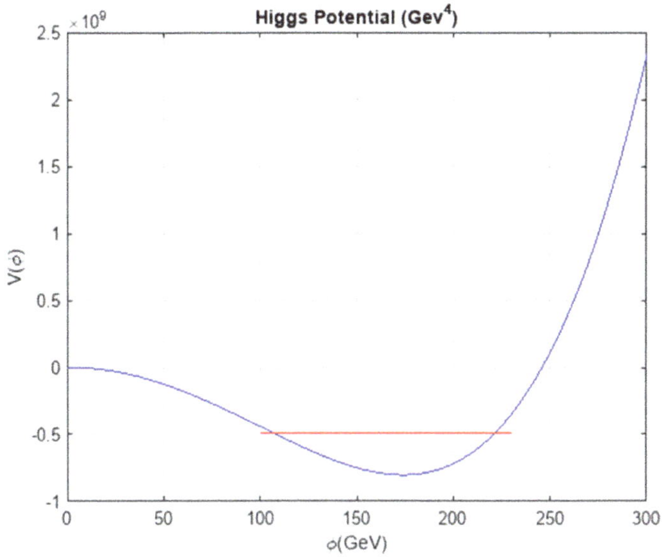

Figure 8.15: Higgs potential as a function of the Higgs field. A minimum occurs at a finite value of the vacuum expectation value. The physical Higgs boson oscillates about that minimum and has a physical mass of 125 GeV.

At present the fundamental particle SM agrees with all present data. However, it has many constants, masses and couplings, which are not predicted, but are experimental inputs. Therefore there is no clear experimental path to explore. Higher energy accelerators may uncover a new and higher mass scale, but the LHC has not yet done so. Thus, that mass scale may require yet another increase in the energy of accelerators. As for cosmology, there is a Standard Model which also agrees with all the present data. However, ordinary matter is only $\sim$5% of all the mass of the Universe. There is dark matter, with no SM candidates in sight and no experimental observations yet beyond that it gravitates. There is also dark energy, at present experimentally compatible with a cosmological constant. It is the majority energy of the Universe, and no candidates exist for it. The Higgs is a SM particle with vacuum quantum numbers, it does not appear to be a dark energy candidate, but it shows that fundamental scalers exist. Hence, there are many, many open questions in physics, and an interesting future seems to be assured.

9 Appendices

References:

No attempt is made in this text to list book or journal references. At present there is a wealth of information out on the web and many excellent search engines. Users are therefore simply asked to do a focused search for their topic of interest using their favorite browser. That appears to the author to be preferable to a long list of texts which are often inaccessible or a list of complex research articles in journals, which themselves might not be open access.

Appendix 9.1 Physics Constants

The MKS system of units is used in the text. When used in the code, the constants are called out in "comments". The table below gives a small subset of the most basic constants. An exception is in the section on cosmology/astrophysics. Because many texts and references for these topics use cgs units they are sometimes employed quoted in that section. The user should be able to easily convert the results to MKS units.

Physics Constants

Quantity	Symbol, equation	Value	Uncertainty (ppb)
speed of light in vacuum	c	$299\ 792\ 458\,\text{m}\,\text{s}^{-1}$	exact*
Planck constant	h	$6.626\ 070\ 040(81) \times 10^{-34}\,\text{J}\,\text{s}$	12
Planck constant, reduced	$\hbar \equiv h/2\pi$	$1.054\ 571\ 800(13) \times 10^{-34}\,\text{J}\,\text{s}$	12
		$= 6.582\ 119\ 514(40) \times 10^{-22}\,\text{MeV}\,\text{s}$	6.1
electron charge magnitude	e	$1.602\ 176\ 6208(98) \times 10^{-19}\,\text{C}$	
		$= 4.803\ 204\ 673(30) \times 10^{-10}$ esu	6.1, 6.1
conversion constant	$\hbar c$	$197.326\ 9788(12)\,\text{MeV}\,\text{fm}$	6.1
conversion constant	$(\hbar c)^2$	$0.389\ 379\ 3656(48)\,\text{GeV}^2$ mbarn	12
electron mass	m_e	$0.510\ 998\ 9461(31)\,\text{MeV}/c^2$	
		$= 9.109\ 383\ 56(11) \times 10^{-31}\,\text{kg}$	6.2, 12
proton mass	m_p	$938.272\ 0813(58)\,\text{MeV}/c^2$	
		$= 1.672\ 621\ 898(21) \times 10^{-27}\,\text{kg}$	6.2, 12
		$= 1.007\ 276\ 466\ 879(91)\text{u}$	
		$= 1836.152\ 673\ 89(17) m_e$	0.090, 0.095
deuteron mass	m_d	$1875.612\ 928(12)\,\text{MeV}/c^2$	6.2
unified atomic mass unit (u)	(mass ^{12}Catom)/12	$931.494\ 0954(57)\,\text{MeV}/c^2$	
	$= (1\,\text{g})/(N_A\,\text{mol})$	$= 1.660\ 539\ 040(20) \times 10^{-27}\,\text{kg}$	6.2, 12

(*Continued*)

Quantity	Symbol, equation	Value	Uncertainty (ppb)
permittivity of free space	$\epsilon_0 = 1/\mu_0 c^2$	$8.854\ 187\ 817\ldots \times 10^{-12}\,\mathrm{F\,m^{-1}}$	exact
permeability of free space	μ_0	$4\pi \times 10^{-7}\,\mathrm{N\,A^{-2}}$ $= 12.566\ 370\ 614\ldots \times 10^{-7}\,\mathrm{N\,A^{-2}}$	exact
fine-structure constant	$\alpha = e^2/4\pi\epsilon_0\hbar c$	$7.297\ 352\ 5664(17) \times 10^{-3}$ $= 1/137.035\ 999\ 139(31)^{\dagger}$	0.23, 0.23
classical electron radius	$r_e = e^2/4\pi\epsilon_0 m_e c^2$	$2.817\ 940\ 3227(19) \times 10^{-15}\,\mathrm{m}$	0.68
(e^- Compton wavelength)/2π	$\lambda_e = h/m_e c = r_e\alpha^{-1}$	$3.861\ 592\ 6764(18) \times 10^{-13}\,\mathrm{m}$	0.45
Bohr radius ($m_{\mathrm{nucleus}} = \infty$)	$a_\infty = 4\pi\epsilon_0 h^2/m_e e^2 = r_e\alpha^{-2}$	$0.529\ 177\ 210\ 67(12) \times 10^{-10}\,\mathrm{m}$	0.23
wavelength of $1\,\mathrm{eV}/c$ particle	$hc/(1\,\mathrm{eV})$	$1.239\ 841\ 9739(76) \times 10^{-6}\,\mathrm{m}$	6.1
Rydberg energy	$hcR_\infty = m_e e^4/2(4\pi\epsilon_0)^2 h^2$ $= m_e c^2\alpha^2/2$	$13.605\ 693\ 009(84)\,\mathrm{eV}$	6.1
Thomson cross section	$\sigma_T = 8\pi r_e^2/3$	$0.665\ 245\ 871\ 58(91)$ barn	1.4

Appendix 9.2 Table of Symbols

The following list is of the symbols used in this text. When invoked in the text, an accompanying definition is made in all cases.

English

A	Angstrom, atomic weight, Area, Ampere
a	semi major axis of ellipse, acceleration, Kerr rotation parameter = J/Mc
$a(t)$	cosmic scale factor
a_o	Bohr radius of hydrogen ground state
B	magnetic field, binding energy
b	semi minor axis of ellipse, impact parameter
c	speed of light
D	diffusion coefficient
$d\Omega$	differential solid angle
e	electron, electronic charge, exponential constant
E	electric field, system energy
eV	electron volt $= 1.6 \times 10^{-19}$ J
F	force
FFT	fast Fourier transform
FH	heat flux
FT	Fourier transform
f	frequency, focal length
G	gravitational constant
g	acceleration of gravity
H	Hubble parameter
h	Planck constant, GR angular momentum $\sim L/c$, scale of metrical distortion
$\hbar$	$= h/2\pi$, reduced Planck constant
I	momenta of inertia, current, wave intensity
J	$= L/m$
j	square root of -1, current density
K	kaon
k	Boltzmann constant, wave number, spring constant
L	length of an object, inductance, angular momentum, luminosity (Watts)

l	angular quantum number, Lagrange density
M	mass of a body, amplitude of drum membrane, CM energy
m	mass of a particle, z angular quantum number
m_e	electron mass
N	total number of objects
N_A	Avogadro's number
n	number density, Bohr quantum number, index of refraction, neutron
P	power, pressure head
p	particle momentum, proton, pressure, dipole moment
Q	charge of a system
q	particle electric charge in units of e
R	radius of an object, reflection coefficient
Rs	Schwarzschild radius
Ruv	curvature tensor
r	coordinate radius, random variable
r_e	classical electron radius
S	Poynting vector
s	arc length, proper time, number of spin states
T	absolute temperature, kinetic energy, period, transmission coefficient, burn time
Tuv	stress-energy tensor
t	time
U	total system energy scaled to mass = E/m or E/kT, binding energy
u	energy density
V	volume, electrical voltage, potential energy
v	velocity
v_T	thermal velocity
x, y	transverse coordinates
Y	nuclear abundance, spherical harmonic
Z	atomic number, impedance
z	longitudinal coordinate

Greek

α	fine structure constant, direction cosine, $a(t)/ao$
β	velocity /c
Γ	reaction rate
γ	photon, ratio of specific heats, SR ratio of energy to mc^2

ε	particle energy, permittivity, dielectric constant
ε_o	vacuum permittivity, hydrogen ground state energy
η	viscosity, photon entropy
θ	polar angle in spherical polar coordinates
κ	atmospheric viscosity, opacity, curvature constant
Λ	cosmological constant
λ	wavelength
$\lambda\!\!\!^{-}$	$= \lambda/2\pi$ reduced Compton wavelength
μ	mean molecular weight, chemical potential
μ_o	vacuum permeability
ν	neutrino
π	pion
ρ	mass, charge density, radius of curvature
ρ_c	critical density
Σ	summation
σ	Gaussian standard deviation, cross section, surface charge density, heat conductivity, Stefan-Boltzmann constant
σ_T	Thomson cross section
τ	period of rotation, process time constant
Φ	magnetic, electric potential
ϕ	azimuthal angle, bend angle in a B field, phase of a wave
ψ	wave function, fluid potential
ω	circular frequency
Ω	ohm, solid angle, fraction of the critical density, rotation frequency
$\sim$	approximately equal to
$\langle\ \rangle$	time average of a quantity

Appendix 9.3 Plot Annotation

An important feature of MATLAB Figures is the ability to annotate them. When a Figure is made, many default options are assumed, and later annotation is often invoked in order to modify the appearance of the Figure such that the user is more satisfied. A surfc plot is an example.

```
[X,Y,Z] = peaks(30);
   surfc(X,Y,Z);
   title('Surface-Contour')
```

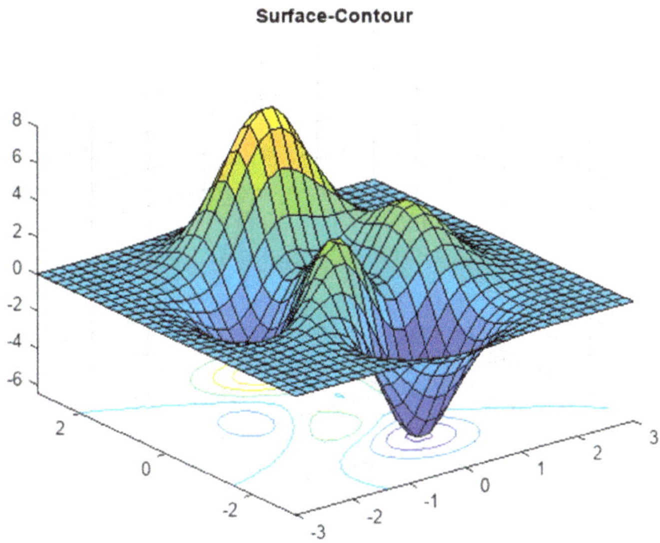

Clicking on the Figure in MATLAB opens the Figure tab in the Live Editor. The axes may be annotated as well as the legend, title, and grid. The size of the display can be mouse controlled. Data points can be examined by hovering over the Figure. The ColorBar can be edited. All these annotations can be done after the execution of the generation of the plot. In this way the user can change all the Figure properties after the fact in order to get the desired display.

Appendix 9.4 Data Fitting

A much more complex fitting procedure than the polynomial fits shown in the body of the text is to use a non-linear chisquare, χ^2, minimization of a set of data points y_i with associated errors σ_i compared to a hypothesized shape which depends on a set of parameters $\alpha, y(x, \alpha)$. The following script is perfectly general and the user can define any set of data points and errors and can define any shape, although changes to the scripts are needed since only a few shapes are predefined. Good starting values for the parameters are needed due to the non-linearity of the procedure. The MATLAB utility "fminsearch" is a general nonlinear minimization utility which is used to do the minimization

$$x^2 = \sum_i^n \{y_i - y(x_i, \alpha)\}^2 / \sigma_i^2$$

The package is exercised using pseudo data which is generated by Monte Carlo methods here distributed as a Gaussian, exponential, power law or a Lorentzian. Which of the predetermined shapes is chosen by the user with an "edit field". The main script is "fit_package". The "data" to be fit can be a set of x values with data y and associated errors, with the number of points also chosen by the user, WT or weight with is the inverse of the square of the r.m.s. of the data. A large error has a small weight in the fit. The fit package is tested by using Monte Carlo distributions to make histograms of pseudo- events with the "data" taken to be the number in the bin (50 bins total) and the error on the data to be the square root of that number. Or the data can be some real data histogrammed in "fit_hst". To validate the package, the "data" are Monte Carlo generated shapes. The error on the fitted parameters are determined by error propagation from the input errors. The errors on the parameters are in general correlated and the full covariance matrix is calculated.

```
%
% close all
clear all
%
% drive the fit_package to check out errors etc.
%
global  X Y Wt Yfit Itype
%
ntot = 10000; % # events, errors scale as sqrt(N)
nb = 50; % bins of histo "data" to fit
%
Itype = 4; % choice of Monte Carlo "data" - Gauss, exp, power law,
Lorentzian
%
for i = 1:ntot
    %
    if Itype == 1
       xmi = 0.0; xmx = 8.0; [xo(i),dum] = Gaus(4, 4, 2);
       % Itype = 1
       ao = [600 4.1 2.1 ]; % starting values for Gaussian,
       peak, mean, sigma
    end
    if(Itype == 2)
       xmi = 0.0; xmx = 5.0; xo(i) = expMC(1,xmi,xmx); % Itype = 2
       ao = [900 1]; % starting values for exp
    end
```

```
    if (Itype == 3)
        %  Power law, cannnot have alf = -1
        xmi = 0.01; xmx = 1.0; alf = -0.5; b = alf+1; % not xmi = 0
        or xmx = 0
        xo(i) = (xmi .^b + rand .*(xmx .^b - xmi .^b)) .^(1.0 ./b);
        ao = [400 -2.1];
    end
    %
    if (Itype == 4)
        % Lorentzian - resonance
        a(1) = 5; a(2) = 1; xmi = 0; xmx = 10; % mean, width, and
        x range
        phmi = (2 .*(xmi - a(1))) ./a(2); phmx = (2 .*(xmx - a(1)))
        ./a(2);
        xo(i) = a(1) + (a(2) ./2) .*tan(atan(phmi) + rand
        .*(atan(phmx) - atan(phmi)));
        a(3) = ntot ./7;
        ao = [5 1 300];
    end
    %
end
%
% for a simple data fit of data yo at xx, set nb  0, simulate it
here
%
[yo, xx] = hist(xo,50);
for i = 1:50
    if yo(i) < 1
        yo(i) = 1; % protect on 0 bins
    end
end
nb = 0;
X = xx; Y = yo; Wt = sqrt(yo); % use globals to directly set up
%
[nxi,erxi,xibin,afit,erra,diag,chs,dof] = fit_package(xo,xmi,
xmx,nb,ao);           % first data with errors plotting then add fit
xplot = linspace(xmi, xmx);
if Itype == 1
    yplot = afit(1); %  peakItype 1 Gaussian
    yplot = yplot .*(exp(-(xplot-afit(2)).^2./(2 .*afit(3)
    .*afit(3))));
end
if Itype == 2
    yplot = afit(1) .*exp(-xplot ./afit(2)); %Itype = 2 exp
end
```

```
%
if Itype == 3
    yplot = afit(1) .*((xplot) .^afit(2)); %Itype = 3 power law
end
if Itype == 4
    yplot = afit(3) ./((xplot - afit(1)) .^2 + (afit(2) ./2) .^2);
    %Itype = 4 Lorentzian
end
figure
plot(xplot,yplot,'-r');
hold('on')
%
fprintf(' Fit parameters, afit = %g\n',afit);
```

```
Fit parameters, afit = 5.00487
Fit parameters, afit = 0.988229
Fit parameters, afit = 332.403
```

```
fprintf(' Parameter errors, dafit = %g\n',diag)
```

```
Parameter errors, dafit = 0.00679583
Parameter errors, dafit = 0.0145562
Parameter errors, dafit = 6.19563
```

```
fprintf(' Chi^2 = %g\n',chs);
```

```
Chi^2 = 57.3247
```

```
fprintf(' DOF = %g\n', dof) ;
```

```
DOF = 47
```

```
errorbar(xibin,nxi,erxi,'ob')
title('Histogram of the Data')
xlabel('x'); ylabel('n(x)')
hold('off')
```

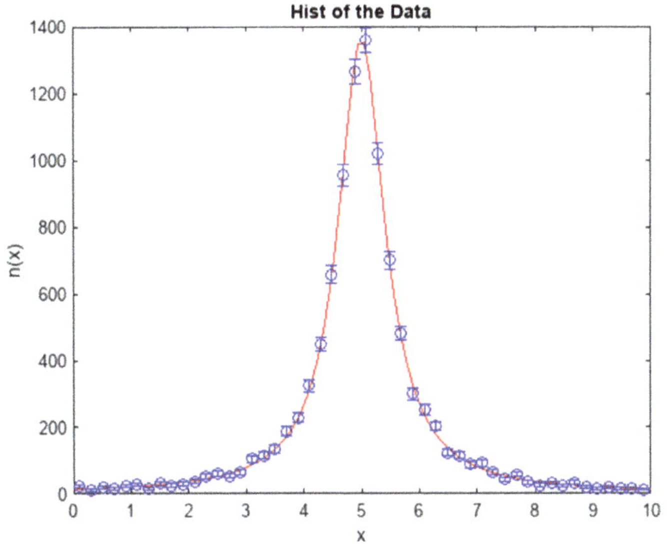


```
function[nxi,erxi,xibin,afit,erra,diag,chs,dof] =  fit_package(xo,
xmi,xmx,nb,ao)
%
% generic fit function specified in function fit_fun
% declare X Y Wt Yfit Itype as global
%
global  X Y Wt Yfit Itype
%
% assume a set of events with a 1-d variable xo
% now make histo of xo from xmi to xmx in nb bins, hist nxi, error
errxi
% bin center xbini - fit to histo points
% or it could be a series of data points with errors - nb = 0
%
if nb == 0
    nxi = Y;
    xibin = X;
    erxi  = Wt;
else
%
    [nxi,erxi,xibin] = fit_hst(xo,xmi,xmx,nb);
end
%
% if nb = 0 the data at xibin is nxi with error erxi, = Y, X, Wt
%
```

```
% input starting values of fit variables a are in ao
% the chisq function to be minimized is defined in fit_fun
%
Y = nxi;
X = xibin;
Wt = 1.0 ./(erxi .^2);
%
% type of fit stored in fit_fun, Itype global variable
%
options = optimset('TolFun',0.1,'TolX',0.1); % set convergence
criteria
afit = fminsearch(@fit_fun,ao) ; % starting values ao
chs = fit_fun(afit); % minimum chisq at best fit
dof = length(Y) - length(afit);
%
% then find err martix on fit values - need explicit deriv vector
of chi
% wrt a. diagonal approximation diag
%
[erra,diag] = fit_err(ao,afit);
end
%
% ------------------------------------------------------------------
%
function[nxi,erxi,xibin] = fit_hst(xo,xmi,xmx,nb)
%
% find histo for fit package
% xo is the vector to be histogranned from (xmi,xmx) - nb bins
% nxi in bin i with error erxi, bin center xibin
%
for k = 1: nb+1
    edg(k) = xmi + ((xmx - xmi) .* (k - 1)) ./nb;
    % edges of hist are xmi and xmx
end
for k = 1: nb
    xibin(k) = edg(k) + (xmx - xmi) ./(2.0 .*nb);
    % bin centers
end
%
nnxi = histc(xo,edg);
for i = 1:nb
    nxi(i) = nnxi(i);
end
erxi = sqrt(nxi);
end
```

```
%
%
-------------------------------------------------------------------
%
% general fit function for least sq fit to measures Y with
weight Wt at
% points X fit parameters a, Yfit a specified function.
%
function[chi] = fit_fun(a)
global X Y Wt Yfit Itype % X Y Yfit Wt Itype are global.
%
% Itype = 1 pure gaussian
% Ityp2 = 2, simple exponential
% Itype = 3, (aX^b) - power law
% Itype = 4, Lorentzian a(1)/[(x-a(2))^2+a(3)^2]\%
%
if Itype == 1
 % a(1) = peak value, a(2) =  mean, a(3) sigma = r.m.s.
 Yfit = a(1) .*(exp(-(X-a(2)).^2./(2 .*a(3) .*a(3))));
 %
end
if Itype == 2
 Yfit = a(1) .*exp(-X ./a(2)); % simple exponential y = a1e(-x/a2)
 a(3) = 0;
end
%
if Itype == 3
 Yfit = a(1) .*(X .^a(2)); % power law a1*(x)^a2
 a(3) = 0;
end
%
if Itype == 4
    % mean and linewidth a1 and a2
    Yfit = a(3) ./((X-a(1)).^2 + (a(2)/2).^2);
    ao = [5 1 1400 ];
end
%
ch = ((Y-Yfit) .^2) .*Wt;
chi = sum(ch);
end
%
% -----------------------------------------------------------------
%
% calculates the error matrix on fit variables a in least sq fit
% companion to fit_fun, get gradient numerically
```

```
%
function[erra,diag] = fit_err(ao,a)
%
% X Y Wt Yfit Itype are assumed global variables, used in
fminsearch, fit_fun
%
global  X Y Wt Yfit Itype
%
% a are best fit params - ao are starting values
%
 npt = length(X);
 na = length(a);
%
% get deriv of the npt Yfit wrt the na parameters, step 1% of
distance to ao
%
  dyda = zeros(na,npt);
  da = abs((a-ao) ./100.0);
  ymin  = Yfit;
%
for i = 1:na
      b = a;
      b(i) = a(i) + da(i); % change 1 at a time
   for j = 1:npt
      dummy = fit_fun(b); % change fron ymin best fit
      dyda(i,j) = (Yfit(j) - ymin(j)) ./da(i);
   end
end
%
% now propagate errors from meas points YY to param a
% assume that the starting errors on YY are uncorrelated - but easy
generalize
%
 covyi = zeros(npt,npt);
 for i = 1:npt
   covyi(i,i) = covyi(i,i) + Wt(i); % diagonal
 end
 covai = dyda * covyi * (dyda'); % off diagonal
 ii = eye(na);
 erra = ii / covai;
```

```
% estimator using diagonal elements
%
 for i=1:na
  diag(i) = sqrt(abs(erra(i,i)));
 end
end
%
```

Appendix 9.5 Solar Properties

Property	Symbol	Value
Radius	R(m)	6.9×10^8
Mass	M(kg)	2.0×10^{30}
Luminosity	L(W)	3.8×10^{26}
Surface temperature	T(°K)	5770
Solar Constant (Earth)	$f(W/m^2)$	1.37×10^3
Schwarzschild radius	R_s(km)	3.0
Number density - mean	$N(1/m^3)$	1.2×10^{23}
Mass density - mean	$\rho(kg/m^3)$	1.4×10^3
Pressure $= GM\rho/R$ - mean	$P(Pa = Nt/m^2)$	3×10^{14}
Core Temperature $= T_c$	T(°K)	1.5×10^7
Core pressure $= P_c$	P(Pa)	2.6×10^{16}
Core density $= \rho_c$	$\rho_c(kg/m^3)$	1.5×10^5
Hydrogen and helium mass fraction, X and Y. $X + Y \sim 1$	X, Y	0.7, 0.28
X and Y, core values, ionized	X, Y	0.35, 0.65
Mean molecular mass fraction, ionized	μ	1.30 H, 0.62

Index

Printed in the USA
CPSIA information can be obtained
at www.ICGtesting.com
JSHW050042210924
69724JS00012B/231

9 789811 295614